Air Quality: Modeling and Assessment

Air Quality: Modeling and Assessment

Editor: Frieda Bush

Callisto Reference
www.callistoreference.com

Callisto Reference,
118-35 Queens Blvd., Suite 400,
Forest Hills, NY 11375, USA

Visit us on the World Wide Web at:
www.callistoreference.com

ISBN: 978-1-64116-187-9 (Hardback)

Printed in China.

Cataloging-in-Publication Data

Air quality : modeling and assessment / edited by Frieda Bush.
p. cm.
Includes bibliographical references and index.
ISBN 978-1-64116-187-9
1. Air quality. 2. Air--Pollution. 3. Air quality management.
4. Air quality--Measurement. 5. Air--Pollution--Measurement. I. Bush, Frieda.

TD883 .A37 2019
628.53--dc23

Table of Contents

Preface

This book has been a concerted effort by a group of academicians, researchers and scientists, who have contributed their research works for the realization of the book. This book has materialized in the wake of emerging advancements and innovations in this field. Therefore, the need of the hour was to compile all the required researches and disseminate the knowledge to a broad spectrum of people comprising of students, researchers and specialists of the field.

Air quality is deteriorated when harmful particulates, biological molecules and gases are present in excessive quantities in the Earth's atmosphere. Such air has a harmful effect on living organisms, food crops and the built environment. In humans, it may lead to diseases, allergies and even death. The sources of air quality degradation or pollution can be either anthropogenic or natural. Anthropogenic sources include smoke released from fossil fuel power stations and motor vehicles, waste deposition in landfills, controlled burning in agriculture, fumes from hair sprays, paints, varnish, etc. Methane released due to the digestion of food in cattle, smoke and carbon dioxide from wildfires, and radon gas generated due to the radioactive decay of radium within the Earth's crust are some of the natural sources of air pollution. All these processes result in the release of harmful pollutants into the atmosphere, which are responsible for air quality deterioration. This book brings forth some of the most innovative concepts and elucidates the unexplored aspects of air quality modeling and assessment. Different approaches, evaluations, methodologies and advanced studies on air quality degradation have been included in this book. With state-of-the-art inputs by acclaimed experts of this field, this book targets students and professionals.

At the end of the preface, I would like to thank the authors for their brilliant chapters and the publisher for guiding us all-through the making of the book till its final stage. Also, I would like to thank my family for providing the support and encouragement throughout my academic career and research projects.

Editor

1

Radon Risk Analysis Through Geostatistical Tools Implemented in a WebGIS

Veronica Distefano, Sandra De Iaco,
Monica Palma and A. Spennato

Additional information is available at the end of the chapter

1. Introduction

Radon (Rn) is a colorless, odorless, tasteless, inert radioactive gas, and derives from the decay of uranium, which is a radioactive element that is found in small quantities in all sediments and rocks.

The International Agency for Research on Cancer (IARC) and the World Health Organization (WHO) classify Rn pollution as the second leading cause of lung cancer after smoking.

Since Rn is present, in the depths of the Earth, in gaseous phase, it reaches the surface because it interacts with other natural elements, such as uranium, thorium and radio (precursor elements); moreover other geo-lithological features, such as the mineralogical composition of the rocks, the underground permeability levels, the presence of faults, fractures and cavities, affect the transport of the Rn on the surface.

In this paper, the spatial distribution of the Rn concentrations in soil gas over a survey area located in the South of Apulian Region (Italy) and its prediction at unsampled points have been discussed. In particular, Ordinary Kriging (OK), Log-Normal Kriging (LK), Cokriging with indicator variable (ICK) and Kriging with Varying Means (KVM) [11, 21, 26, 27] have been used to predict Rn concentrations over the study area.

In this context, the integration of a Geographical Information System (GIS) and geostatistical tools can certainly support the evaluation of alternative scenarios, possible strategies for a sustainable development [1].

In addition, the geoprocessing tools (e.g., buffering, overlay, union/intersection, interpolation) can be used to combine and interpret data obtained from different sources. The construction of a GIS project facilitates the data sharing and the integration of environmental and demographic data, as well as the results of spatial analysis.

Thus, the usefulness of geostatistical techniques for monitoring Rn concentrations in soil gas and the integration of these results in an *ad hoc* WebGIS, named RnWebGIS, have been described. This innovative tool offers dynamic scenarios for monitoring and analyzing different environmental variables [7, 39]. The proposed RnWebGIS is based on an Open Source Software, which enables users to get online spatial and environmental information.

Hence, after a discussion on the integration of GIS and geostatistical tools (Section 2), a brief review of some geostatistical techniques for modeling and prediction purposes has been presented (Section 3). A thorough case study concerning the Rn concentration in soil gas measured in several sites over Lecce district has been developed (Section 4). Finally, the RnWebGIS, where all the available geo-lithological information of the area under study, as well as the Rn prediction results have been integrated, has been proposed and the steps for its implementation have been detailed.

2. GIS and geostatistical tools

Environmental risk management involves the integrated use of geostatistical techniques and several tools provided in a GIS.

A GIS is a repository of tools for capturing, storing, updating, manipulating, analyzing, and displaying all geographic information [9]. The data are stored as collections of relational tables logically associated with each other by shared attributes. On the other hand, Geostatistics provides advanced techniques to predict a variable of interest at unsampled points and to support decisions concerning monitoring, sampling, planning and re-qualification of a territory [10].

Fig. 1 shows a flowchart regarding the steps which characterize the connection between GIS and geostatistical tools.

The first step is based on the creation of a geodatabase, that is a relational database which stores a series of tables holding feature classes, raster datasets, attributes and geographic information represented by shapefiles (i.e. geospatial vector data format which can spatially describe vector features, such as polygon, line or point, with defined shape and geometric location). Then, the next step concerns the georeferencing, that is the process of assigning real-world coordinates to each pixel of the raster data. In this case, these coordinates are often obtained through field surveys, by collecting coordinates with a GPS device, or by using Google Earth software. On the other hand, the shapefile data do not need to be georeferenced, but simply loaded in a geodatabase by specifying the coordinate system associated to the datasets; hence, the vector features in the shapefile will be represented by using the given coordinate system.

The third step involves the implementation of the geostatistical analysis which is organized in various phases, such as the exploratory spatial data analysis (i.e. mapping the spatial distribution using thematic maps), the estimation and modeling the spatial correlation structure exhibited by the data, predicting the variable of interest at unsampled locations by using spatial interpolation techniques (such as Simple, Ordinary and Universal Kriging, Indicator Kriging, Cokriging) and computing the corresponding prediction errors. Finally, the last step consists in the visualization of geostatistical results by producing reliable prediction maps, probability maps, and other useful cartographic representations.

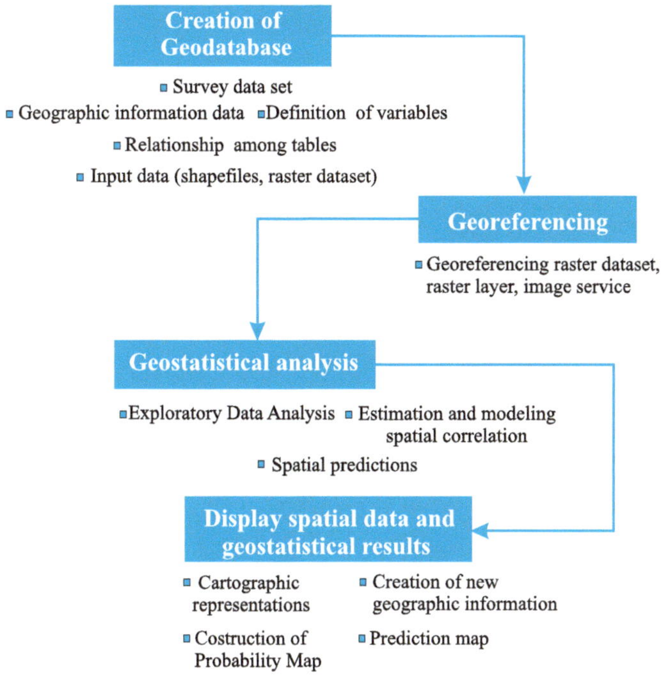

Figure 1. Integration between GIS and geostatistical tools.

The integration between GIS and Geostatistics has been of interest since the early 1990s, when [20] and [8] illustrated the potential benefits of a close link between information systems and Spatial Analysis. Many contributions, concerning the integration of spatial statistical methods and a GIS environment, have been proposed in the literature [2, 3, 6, 9, 14, 37] over the years.

In order to improve the integration between GIS and Geostatistics, different methods of spatial analysis have been implemented in several kinds of GIS softwares [1, 3]. For example, among the most widely used commercial GIS packages, it is worth citing *ArcGIS*, *Geomedia* and *Mapinfo*, which offer extensions for geostatistical analysis; on the other hand, among the *Open Source* packages it is worth mentioning *GRASS* and *Quantum GIS* which use R, Java and Python in order to implement the geostatistical tools. But, the collection of spatial data analysis softwares now available does not completely include all the computational aspects. Therefore, in order to satisfy the specific requirements of the user it is often necessary to develop *ad hoc routines* for sharing GIS data, implementing macros and customized libraries of geostatistical functions in GIS environment, creating suitable script [36].

3. Geostatistical basic notions

Geostatistical methods [15, 23, 28] are useful applied in many areas of environmental sciences (Geography, Geology, Environmental Sciences, Ecology, Meteorology, Agronomy), for studying phenomena which vary continuously over the survey domain. Usually, the final aim of geostatistical analysis consists of obtaining predictions of the variable of interest at unsampled points of the study area. In this case, the use of stochastic models to describe the spatial behaviour of the phenomenon under study, as well as its spatial uncertainty, is reasonable.

In the following sections, the most applied geostatistical techniques, developed for modeling and prediction purposes, both in the univariate and multivariate case, will be briefly described.

3.1. Modeling and prediction in the univariate case

In Geostatistics, the available spatial data concerning the measurements of a single variable, at n points located over the area of interest, are considered as a particular realization of the random variable $Z(\mathbf{u}_\alpha), \alpha = 1, 2, \ldots, n$, where $\mathbf{u} \in D \subseteq R^d$, $d \in N_+$. This random variable is called in [25] *regionalized variable*, in order to highlight the feature of the spatial dependency for the variable itself. The spatial distribution of the random variable Z is usually interpreted as the result of two overlapping structures: a systematic structure at the macroscopic level, which describes the global features of the same natural process, and another random structure at the microscopic level, which represents the local irregularity of Z.

In particular, by assuming the existence of the first and the second-order moments of the random variable Z, the basic stochastic model for Z is defined as follows:

$$Z(\mathbf{u}) = m(\mathbf{u}) + Y(\mathbf{u}), \qquad \mathbf{u} \in D \subseteq R^d,\, d \in N_+, \tag{1}$$

where $m(\mathbf{u})$, usually called *trend*, describes the large-scale variations and $Y(\mathbf{u})$ is a second-order stationary random field which represents the local fluctuations of Z.

Under the second-order stationary hypotheses, the expected value of Z exists and does not depend on $\mathbf{u} \in D$; moreover the variogram exists and can be defined as follows:

$$\gamma(\mathbf{h}) = 0.5\, E[Z(\mathbf{u}) - Z(\mathbf{u} + \mathbf{h})]^2, \tag{2}$$

where $\mathbf{h} = \mathbf{u} - \mathbf{u}'$ is the separation vector between two support points $\mathbf{u}$ and $\mathbf{u}'$ of the spatial domain D.

The variogram, by definition, is a measure of dissimilarity, in the sense that it increases when the difference between $Z(\mathbf{u})$ and $Z(\mathbf{u} + \mathbf{h})$ increases; thus, it is usually characterized by small values for short spatial distances.

After computing the sample variogram [32], as a measure of the spatial correlation of the variable of interest, a theoretical admissible model has to be fitted to the estimated variogram.

It is worth noting that the condition for a function to be a variogram is that it be conditionally negative definite [11]. However, given the difficulties to verify this condition, it is advisable for users to look for the best model among the wide parametric families whose members are known to be conditionally negative definite functions. This condition ensures a unique solution in the kriging system (system of linear equations based on the variogram matrix of observations in different spatial points of the domain, as it will be discussed hereafter) used to tackle the general problem of predicting the process of interest, starting from the observed values.

Various classes of variogram models have been proposed in the literature; however the selection of an appropriate class of models is in general based on its geometric features and theoretical properties [10, 12, 13].

Once the spatial correlation of the variable under study has been properly modelled, some well-known based-variogram prediction techniques can be used in order to estimate the analyzed variable at unsampled points over the domain of interest.

The most applied spatial predictor is the Ordinary Kriging (OK) [10, 23] which is based on the following estimator:

$$\widehat{Z}^{OK}(\mathbf{u}) = \sum_{\alpha=1}^{n} \lambda_{\alpha}^{OK}(\mathbf{u}) Z(\mathbf{u}_{\alpha}), \tag{3}$$

where the kriging weights λ_{α}^{OK}, $\alpha = 1, 2, \ldots, n$, depend on the variogram model and the sampling configuration.

The predictor $\widehat{Z}^{OK}(\mathbf{u})$ satisfies both the unbiasedness and the efficiency condition.

In particular, the first property is satisfied if the weights λ_{α}^{OK}, $\alpha = 1, 2, \ldots, n$, of the linear combination are such that

$$\sum_{i=1}^{n} \lambda_{\alpha}^{OK}(\mathbf{u}) = 1. \tag{4}$$

On the other hand, $\widehat{Z}^{OK}(\mathbf{u})$ is efficient, if the weights are chosen in order to minimize the error variance $\sigma_E^2 = E[R(\mathbf{u})]^2 = E[\widehat{Z}(\mathbf{u}) - Z(\mathbf{u})]^2$, under the unbiasedness condition.

Finally, it is important to point out that, if the variable under study is characterized by a high coefficient of variation and/or by different levels of variability, traditional prediction methods, such as OK, might not be appropriate. In particular, when the experimental data shown a skewed distribution, it is possible to apply a non-linear transformation at the observed data, such that the distribution of the transformed data is symmetric. In many cases, the logarithmic transformation is useful for this aim. Hence, structural analysis and kriging are performed by considering the log-transformed data.

3.2. Modeling and prediction in the multivariate case

In many environmental applications, the phenomenon under study is the result of the simultaneous behavior of several variables, which are correlated each other and measured at some points of the area of interest.

In this case, a multivariate spatial structure [34] characterizes the observed data which can be considered as a realization of a multivariate random field (*MRF*) $\{\mathbf{Z}(\mathbf{u}),\ \mathbf{u} \in D \subseteq \mathbb{R}^d\}$, with

$$\mathbf{Z}(\mathbf{u}) = [Z_1(\mathbf{u}), Z_2(\mathbf{u}), \ldots, Z_p(\mathbf{u})]^T, \tag{5}$$

where $p \geq 2$.

Under second-order stationarity, the mean vector of $\mathbf{Z}$ exists and does not depend on $\mathbf{u}$, i.e.:

$$E[\mathbf{Z}(\mathbf{u})] = \mathbf{M} = [m_1, m_2, \ldots, m_p]^T, \qquad \mathbf{u} \in D, \tag{6}$$

and the $(p \times p)$ variogram matrix $\mathbf{\Gamma}$ exists and depends on the spatial separation vector $\mathbf{h}$ between two support points ($\mathbf{u}$ and $\mathbf{u}'$), i.e.:

$$\mathbf{\Gamma}[\mathbf{Z}(\mathbf{u}), \mathbf{Z}(\mathbf{u}')] = E\left\{[\mathbf{Z}(\mathbf{u}) - \mathbf{Z}(\mathbf{u}')][\mathbf{Z}(\mathbf{u}) - \mathbf{Z}(\mathbf{u}')]^T\right\} =$$

$$= \mathbf{\Gamma}(\mathbf{h}) = [\gamma_{ij}(\mathbf{h})], \tag{7}$$

where $\gamma_{ij}(\mathbf{h}),\ i,j = 1,2,\ldots,p,$ is the cross-variogram between $Z_i(\mathbf{u})$ and $Z_j(\mathbf{u}+\mathbf{h})$, when $i \neq j$, and the direct variogram of the $Z_i(\mathbf{u})$, when $i = j$.

In multivariate Geostatistics, *Ordinary Cokriging* (OCK) [34] is a well-known technique used to obtain prediction of one or all the variables under study, at unsampled points of the spatial domain. This technique considers the following linear predictor of the spatial random vector defined in (5):

$$\widehat{\mathbf{Z}}^{OC}(\mathbf{u}) = \sum_{\alpha=1}^{n} \mathbf{\Lambda}_{\alpha}^{OC}(\mathbf{u})\mathbf{Z}(\mathbf{u}_{\alpha}), \tag{8}$$

where $\mathbf{u} \in D$ is an unsampled point in the domain D, $\mathbf{u}_{\alpha} \in D$, $\alpha = 1,2,\ldots,n$, are the sampled points in the domain D and $\mathbf{\Lambda}_{\alpha}^{OC}(\mathbf{u})$ are the $(p \times p)$ matrices of weights, which are determined by ensuring unbiasedness and efficiency conditions.

The cokriging predictor requires a model for the variogram matrix (7). The *Linear Model of Coregionalization* (*LMC*), widely discussed in [23] and [34], is the most commonly model used in the applications. In particular, the *LMC* can be written as

$$\mathbf{\Gamma}(\mathbf{h}) = \sum_{l=1}^{L} \mathbf{B}_l\, g_l(\mathbf{h}), \tag{9}$$

where $\mathbf{B}_l$, $l = 1,2,\ldots,L$, are $(p \times p)$ positive definite matrices, called *coregionalization matrices*, while $g_l(\mathbf{h})$, $l = 1,2,\ldots,L$, are basic variograms which correspond to different scales of spatial variability.

A particular case of cokriging, which allows non-parametric, non-linear predictions [22], is the *Cokriging with Indicator Variable* (ICK), where the values observed for one and/or all the variables under study, are replaced by indicator values (that is 1 or 0, depending on a condition specified by the analyst which is of interest for the phenomenon under study).

Another very useful prediction technique, developed in the context of multivariate Geostatistics, is the *Kriging with Varying Means* (KVM), which corresponds to an extension of the *simple kriging* used in the univariate case. In the context of two or more variables, the stationary mean of the primary variable is replaced with a known varying mean according to the secondary information [21]. Hence, the KVM predictor is computed as follows:

$$\widehat{Z}^{KVM}(\mathbf{u}) = \sum_{\alpha=1}^{n} \lambda_{\alpha}^{KVM}(\mathbf{u})[Z(\mathbf{u}_{\alpha}) - m^{KVM}(\mathbf{u}_{\alpha})] \tag{10}$$

where $\mathbf{u} \in D$ is the unsampled point, while λ_{α}^{KVM}, $\alpha = 1, 2, \ldots, n$, are the weights of the predictor. Note that if the secondary variable is categorical, the primary local mean is the mean of the primary variable within a specific category of the secondary variable.

In the proposed case study, the above mentioned prediction techniques will be performed and the main prediction results will be discussed.

4. Case study: the survey area and the data set

The study area is located in the Southern part of Italy (Lecce district in the Apulian region), at 199 meters above the sea level (Fig. 2).

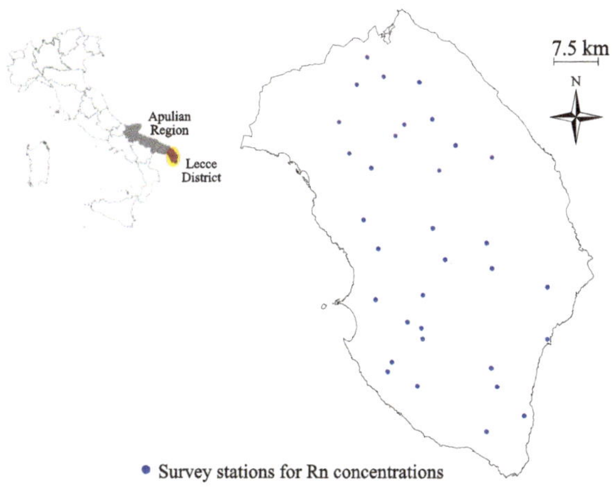

Figure 2. Location map of survey stations in Lecce district.

It is characterized by a limestone substrate and the presence of sink-holes, caves and underground drainages. Moreover, the area is characterized by 6 different lithotypes, with predominance of sandstone (45.7%), limestone and dolomitic limestone (29.4%), as well as siltstone and sandstone (23.9%). Because of this particular lithology, a medium-high level of underground permeability marks the area out. Fig. 3 illustrates the maps of geo-lithological features and permeability levels.

The available geo-lithological information about the survey area represents the *soft* information which can better explain the outdoor Rn concentrations and the corresponding exhalation rate from soil. In order to use the soft information in the modeling and prediction stages of the geostatistical analysis, a regular georeferenced grid covering the area under study (Fig. 4-a)) has been defined. The node separation distance (along the horizontal and vertical directions) has been set equal to 2 km. Thus, the soft data

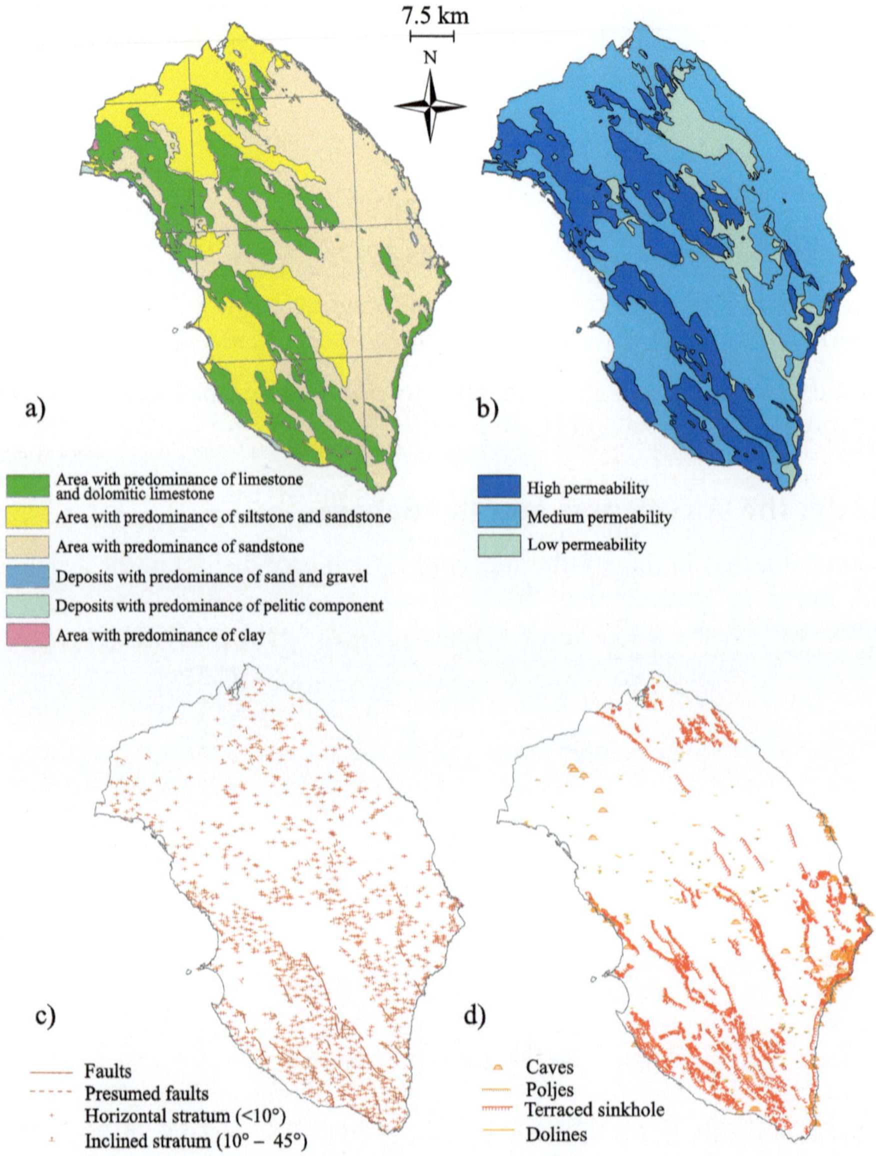

Figure 3. Maps of geo-lithological features in Lecce district. a) Map of geo-lithological features. b) Map of underground permeability levels. c) Map of tectonics features. d) Map of karst forms.

have been stored in a GIS project by using Arcmap of the ArcGIS software and the shapefile (data sources: http://www.sit.puglia.it) concerning the geo-lithological features, the underground permeability level, the tectonic (such as distance from tectonic elements) and karst features (such as distance from karst forms). In particular, by using the *spatial join* tool (a geoprocessing tool of Arcmap), all the soft data have been assigned to each grid point (for example, for each grid point it is possible to identify the type of geo-lithology composition, the distance grid point-faults, the distance grid point-karst forms).

On the other hand, the available Rn concentrations in soil gas have been considered as *hard* information. In particular, the in soil gas Rn levels have been measured at 32 locations over

Lecce district (Fig. 2), during a campaign conducted in June-July 2012 on the basis of a suitable spatial sampling.

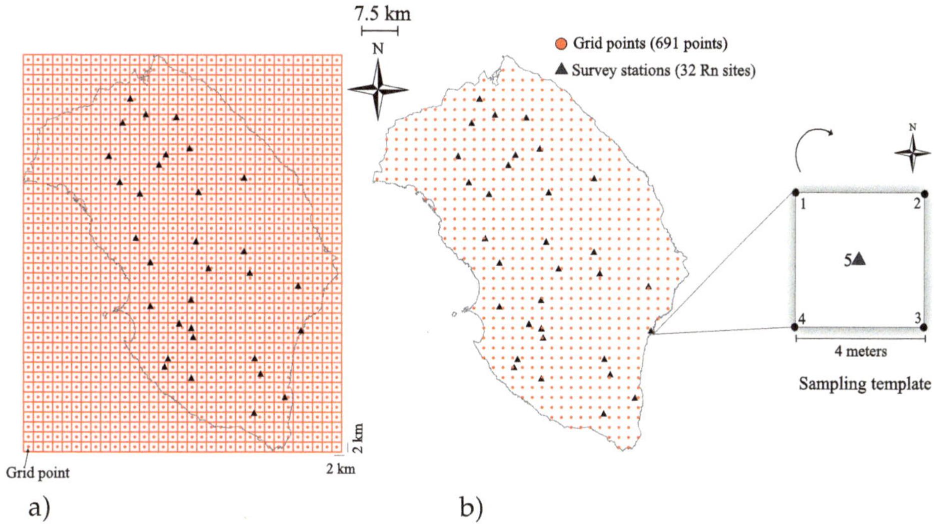

Figure 4. a) Regular grid of cell size 2 × 2 km and grid points with node separation distance equal to 2 km. b) Survey stations for Rn concentrations in soil gas, together with the sampling template.

For each selected site, the Rn concentrations have been detected by using an active sampling-based system, in which the Rn and its decay products are conveyed in proximity of the radiation detector through a mechanical pumping. A special instrument (AlphaGUARD radon monitor by Saphymo GmbH), equipped with an ionization chamber has been used. The chamber has been connected to a probe, introduced into the ground up to a maximum of 1 meter deep, and to a pump for the aspiration of Rn gas at fixed intervals of time. In particular, at each sample location, the measurements of the Rn levels in soil gas have been performed, every minute for 20 minutes, by aspirating the air through the probe introduced into the ground at 5 points according to a regular template (sampling template in Fig. 4). The template consists of a square with a side of 4 meters; thus the Rn measurements have been made both in the center and the corresponding vertices. Hence, both the center and the vertices of the square have been numbered, starting from the top of the North-West and proceeding clockwise, as considered in the sampling protocol.

Finally, the average of the observed Rn concentrations for each site has been assigned and georeferenced at the central point of the sampling template (number 5 in Fig. 4).

All the 32 sample values (hard data) have been georeferenced and stored in a GIS project; then they have been integrated with the soft information.

4.1. Geostatistical analysis: main results

In the following the spatial concentrations of the Rn in soil gas have been analyzed and predicted at unsampled points of the study area by applying different techniques, such as OK, LK, ICK and KVM.

Note that, the above mentioned multivariate interpolation methods utilize both image-derived (soft) data which are more exhaustive and higher accurate than hard data, typically sparse over the domain. Using ICK and KVM is justified when the measurements of environmental variables can be characterized by a high coefficient of variation and/or by different level of variability; indeed in such cases traditional prediction methods, such as OK and LK, are not appropriate because they do not account for these features.

Ordinary kriging predictions

In order to predict Rn concentrations in soil gas by using OK method, the structural analysis has been performed on the measured values. After computing the sample variogram, the following Gaussian variogram model has been fitted:

$$\gamma(\mathbf{h})^{OK} = 80 + 263 \left[1 - \exp\left(-3||\mathbf{h}||/20\right)^2\right] \quad (11)$$

where nugget and sill correspond to 80 and 263, respectively, while range is equal to 20 km. Fig. 5-b) shows the estimated Rn concentrations in soil gas obtained by using OK.

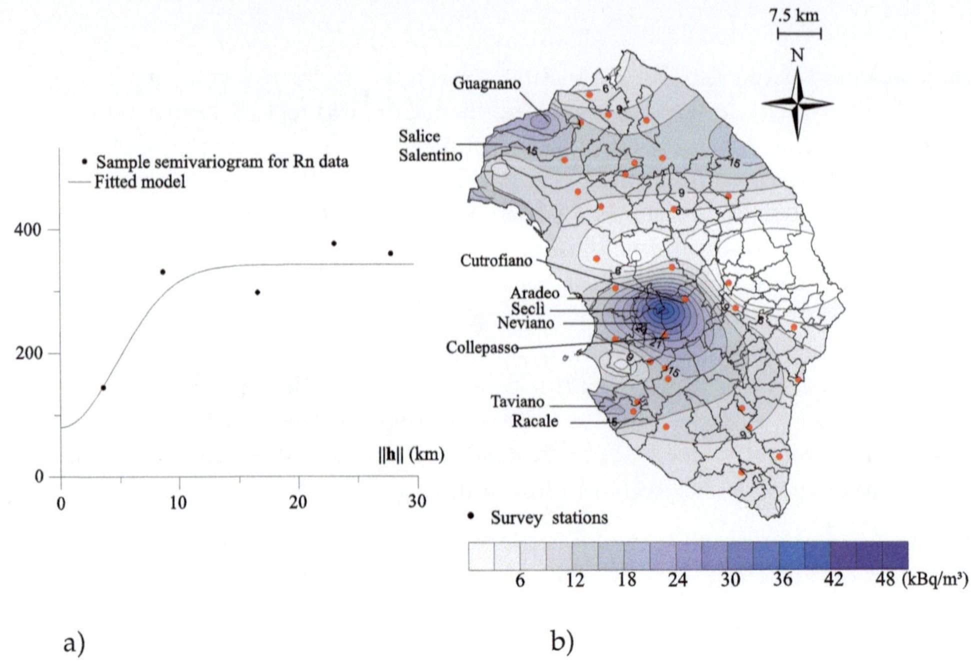

Figure 5. a) Sample variogram and fitted model for Rn concentrations. b) Contour map of the estimated Rn concentrations using Ordinary kriging.

In particular, the prediction map in Fig. 5-b) highlights three Rn prone area, characterized by high Rn concentration values, in the North-Western, South-West and South-South-West of the domain of interest.

Log-Normal kriging predictions

In this case, the Rn measured values have been first processed by applying the logarithmic transformation. Hence, the structural analysis has been performed on the log-transformed data. In Fig. 6-a) the sample variogram computed for the log-transformed data is illustrated together with the fitted model. In particular, the variogram model considered for the LK prediction technique is given below:

$$\gamma(\mathbf{h})^{LK} = 0.1 + 1.2\left[1 - \exp\left(-3||\mathbf{h}||/20\right)^2\right] \tag{12}$$

where nugget and sill correspond to 0.1 and 1.2, respectively, while range is equal to 20 km.

Fig. 6-b) shows the predictions for Rn concentrations obtaining by using LK, after a back-transformation on the data has been applied.

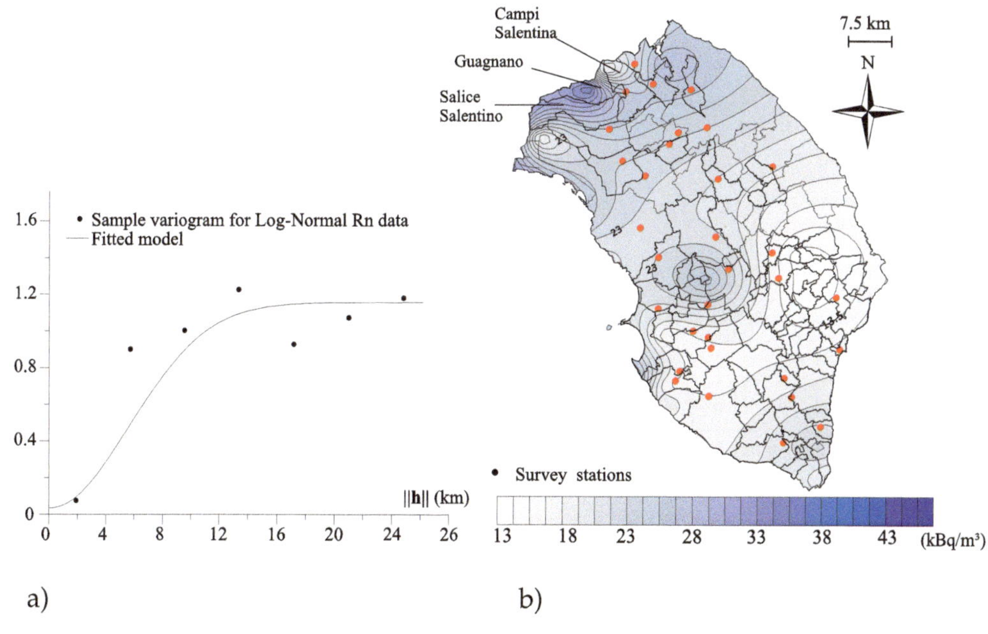

Figure 6. a) Sample variogram and fitted model for Log-Normal Rn concentrations. b) Contour map of estimated Rn concentrations using LK, after a back-transformation on the data.

Despite the previous case (Fig. 5-b)), the prediction map illustrated in Fig. 6-b) highlights only one area with high Rn concentrations, located in the North-Western part of the domain of interest. From the prediction LK results, it is evident the significant *smoothing effect* produced by the interpolation process based on LK.

Cokriging predictions with indicator variable

This method has been applied by using a double source of information, i.e. the sampled Rn concentrations (primary variable) and the underground permeability levels (secondary or auxiliary variable). Evidently the auxiliary information is richer over the study area than the primary variable, hence it has been used to improve predictions of the Rn concentrations in soil gas. In particular, the auxiliary data, which represent the three permeability classes for any node of the grid, have been derived by the software Arcmap of ArcGIS and the shapefile concerning the underground permeability level in Lecce district. A value has been assigned to each point of the grid and to each survey station, according to the different permeability levels (the value 1 for low permeability, the value 2 for medium permeability and the value 3 for high level of permeability).

In this case, the following *LMC* [33, 34] has been fitted to the sample variogram matrix:

$$\mathbf{\Gamma} = \mathbf{B}_1\, g_1(\mathbf{h}) + \mathbf{B}_2\, g_2(\mathbf{h}) \tag{13}$$

where g_1 is a nugget model and g_2 is a gaussian model [23] with unit sill and range equal to 20 km; while $\mathbf{B}_l, l = 1, 2$, are the following coregionalization matrices:

$$\mathbf{B}_1 = \begin{bmatrix} 80 & 2 \\ 2 & 0.125 \end{bmatrix}, \qquad \mathbf{B}_2 = \begin{bmatrix} 263 & 4.8 \\ 4.8 & 0.088 \end{bmatrix}. \tag{14}$$

The sample direct and cross variograms, shown in Fig. 7-a), have been modelled as follows:

$$\gamma_{11}(\mathbf{h}) = 80 g_1(\mathbf{h}) + 263 g_2(\mathbf{h}),$$

$$\gamma_{22}(\mathbf{h}) = 0.125 g_1(\mathbf{h}) + 0.088 g_2(\mathbf{h}),$$

$$\gamma_{12}(\mathbf{h}) = 2 g_1(\mathbf{h}) + 4.8 g_2(\mathbf{h}).$$

Finally, cokriging predictions of Rn concentrations in soil gas have been obtained as shown in Fig. 7-b).

In particular, the prediction map shown in Fig. 7-b) highlights three Rn prone areas, characterized by high Rn concentration values, in the North-Western, South-West and South-South-West of the domain of interest.

Kriging predictions with Varying Means

In order to apply this interpolation technique to the available data set, the Rn concentrations have been grouped by the three permeability classes, then the hypothesis of the presence of different mean levels of Rn concentrations in soil gas has been assumed.

After the structural analysis, the following variogram model (Fig. 8-a)) has been considered:

$$\gamma(\mathbf{h})^{KVM} = 170\ [1 - \exp\left(-3||\mathbf{h}||/15\right)] \tag{15}$$

where sill is equal to 170 and range equals to 15 km.

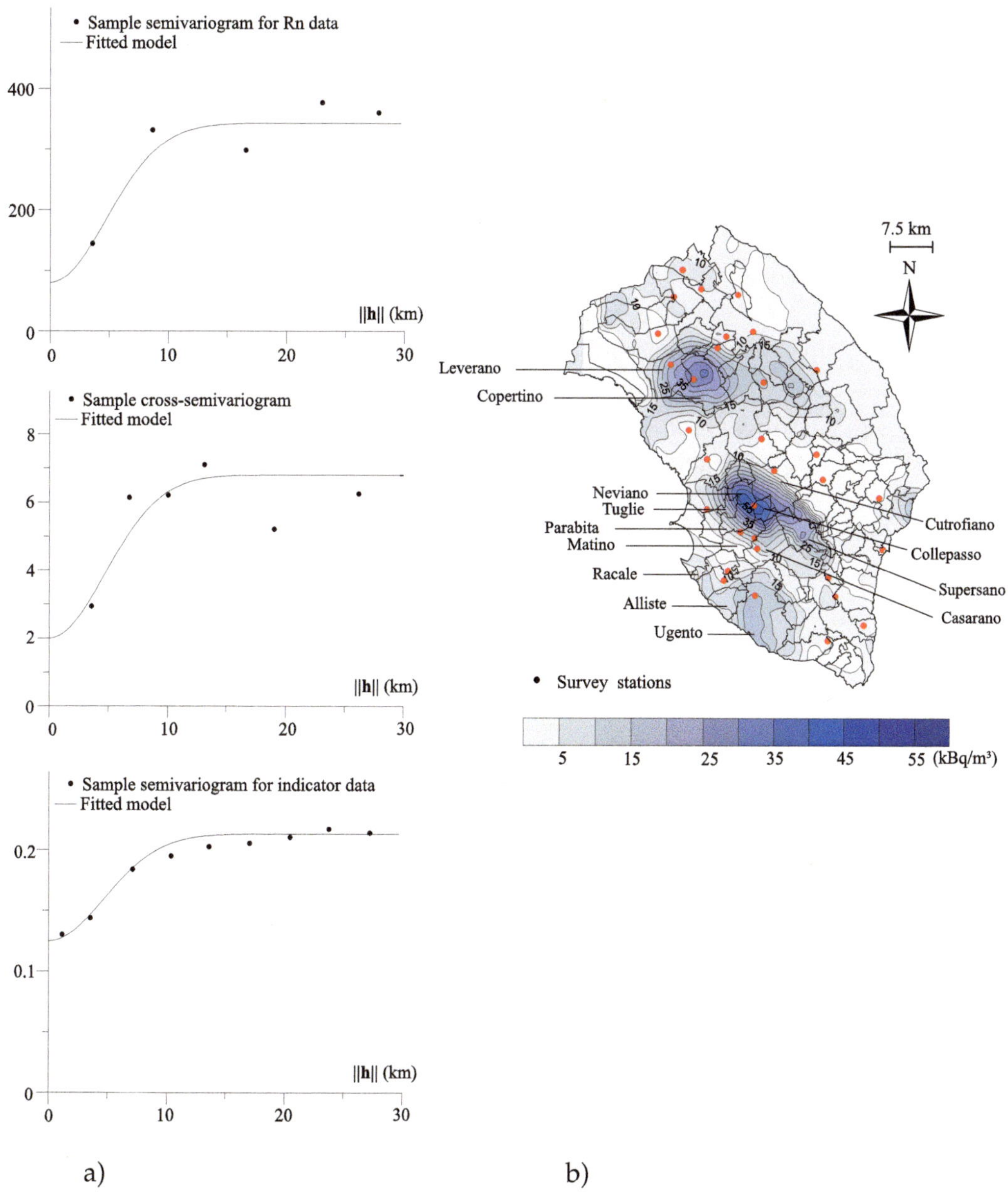

Figure 7. a) Sample direct and cross variograms and fitted model for Rn and indicator data. b) Contour map of estimated Rn concentrations using cokriging with indicator variable.

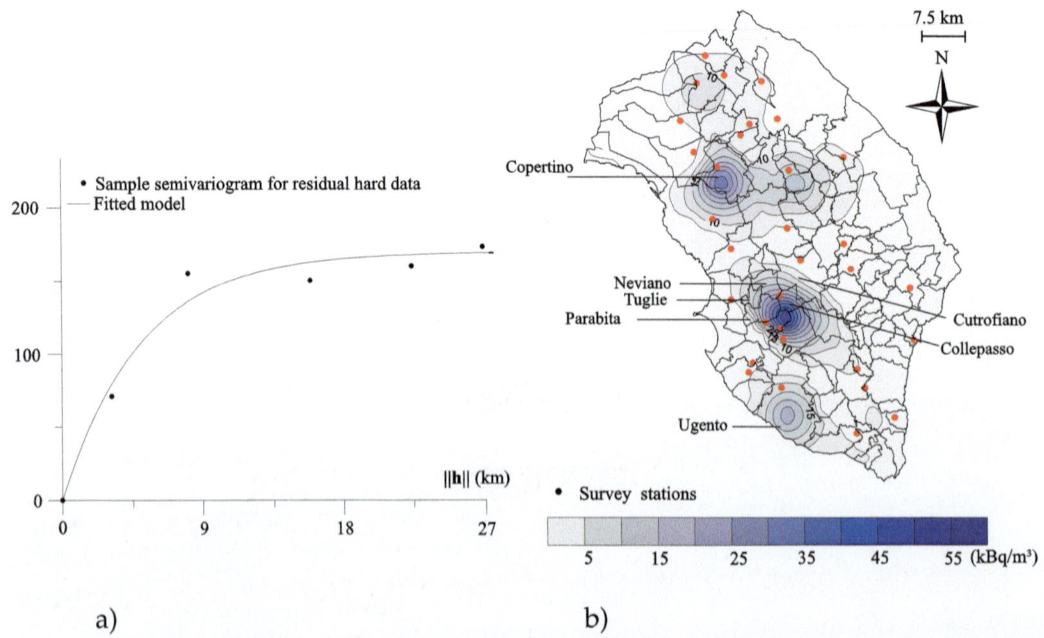

Figure 8. a) Sample variogram and fitted model for residual hard data. b) Contour map of estimated Rn concentrations using KVM.

Hence, predictions for Rn concentrations in soil gas have been obtained for the study area (Fig. 8-b)) by using KVM.

In particular, the prediction map in Fig. 8-b) confirms that high Rn concentration values are located in the North-Western, South-West and South-South-West of the domain of interest.

4.2. A comparison among the prediction results

All the four contour maps illustrated in Figg. 5-b)-8-b) reproduce almost a similar behaviour, with high levels of Rn concentrations located in the Western part of the domain of interest. However, the predictions maps obtained from the KVM and ICK techniques, exploit the auxiliary information and identify the Rn prone areas more accurately than the ones computed by using the OK and LK techniques.

Moreover, the four interpolation techniques have highlighted the existence of a direction (135° from North to South, close to the Western coast of the study area), along with the Rn concentrations in soil gas present high levels.

The accuracy of the predictions has been measured through the *Root Mean Square Error* (RMSE) index [35, 38], defined as follows:

$$RMSE = \sqrt{\frac{1}{n}\sum_{\alpha=1}^{n}[z(\mathbf{u}_\alpha) - \widehat{z}(\mathbf{u}_\alpha)]^2}, \tag{16}$$

where $z(\mathbf{u}_\alpha)$ represent the measured values, $\widehat{z}(\mathbf{u}_\alpha)$ correspond to the estimated values and n is the number of sample data.
By analyzing the RMSE shown in Fig. 9 it is evident that the accuracy of predictions is lower for OK and LK than the other two interpolation techniques.

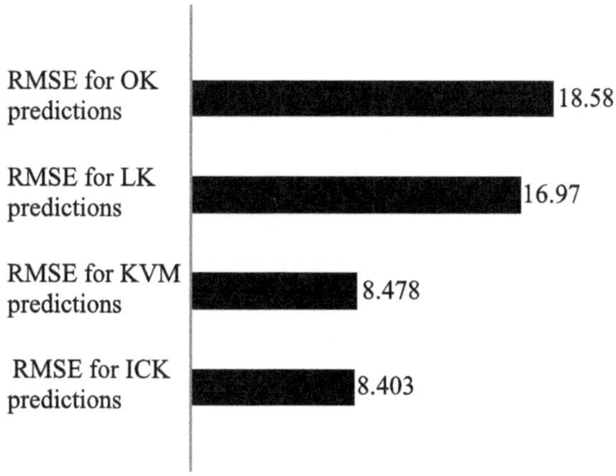

Figure 9. RMSE computed for the OK, LK, KVM and ICK predictions.

However, if the comparison is expressed in terms of the *relative percentage changes* of RMSE index:

- the use of LK method with respect to the OK, improves the predictions reliability of 9.5%;
- the use of ICK method with respect to the KVM, doesn't produce significant improvements for the predictions accuracy (just only 1%);
- the use of multivariate methods (such as KVM and ICK), improve the predictions reliability by about 51%, with respect to LK, and about 55% with respect to OK.

The better performance of the multivariate prediction techniques, compared with the univariate ones (OK and LK), is due to the appropriate use of the information given by the auxiliary variables (soft data). Moreover, among the univariate prediction techniques, the LK has shown a better performance since the variable under study has a skewed distribution.

5. Towards the implementation of a RnWebGIS

The evolution of information technology and GIS allow a potentially unlimited number of users to access and manage geospatial information online [18, 19].

A WebGIS offers an efficient way to provide geospatial information and Web Services to concurrent users without installing a GIS. For this reason a WebGIS system is defined as a web information system that provides geographic information on the web through the Hypertext Transfer Protocol (HTTP) and the HyperText Markup Language (HTML) [17].

Therefore, a WebGIS is different from other web maps, such as Google Earth or Google Maps, since it integrates Internet/web technologies with functionalities of a GIS structure [7].

Fig. 10 shows a schematic diagram regarding the WebGIS architecture, with the following essential steps:

- first step: a user sends a request to the *web browser*;
- second step: such a request is sent via Internet and processed by the server. Note that the WebGIS systems are equipped with a server mapping, such as *MapServer* [24] that obtains information from the geographic databases;
- third step: the request is displayed to the user through the *web browser*.

In the above-mentioned architecture the responsibility is on the server side, while the user (or client) needs only a browser in order to utilize the WebGIS and WebGIS applications. Therefore, the client sends requests to the GIS server which processes it and sends it to the client. Hence, the server performs the task of a *file server*, which contains the mapping data included in the WebGIS, while the client visualizes and consults the cartographic map. Therefore, the main elements of a WebGIS are the following: a) a web server, b) a scripting language, c) a database, d) a server mapping and e) an interface for the server mapping.

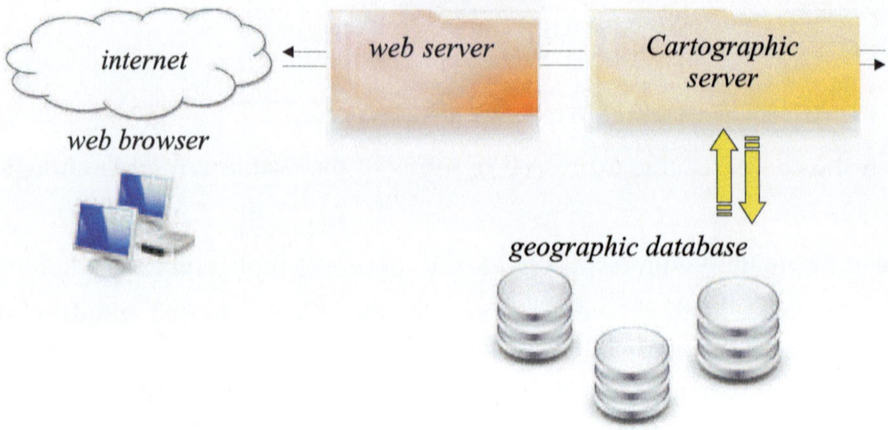

Figure 10. Schematic diagram of the WebGIS architecture.

In this paper, the RnWebGIS has been realized by using Open-source technologies, such as the *web server* Apache, a *MapServer* Project and the client interface *Pmapper*.

In the proposed RnWebGIS, the following databases have been included, such as:

- administrative boundaries:
  - the provincial borders;
  - the municipal boundaries;
  - the urban center boundaries;
- Rn concentrations and the related geographical information (Rn sites);
- the cartographic maps of the permeability and the faults in Lecce district;
- the Web Map Service (WMS), which allows to generate the satellite map of the Lecce district;
- the geostatistical results, which includes the prediction map of ICK and the level curves for the Rn concentrations.

It is important to highlight that the obtained RnWebGIS allows the visualization of alphanumeric information associated with geographic features of interest, through the use of appropriate tables, or to implement the connection to other *web* pages or *database*, by using suitable *hyperlinks* functions.

Through specific queries on thematic maps, the user can display, in tabular form, the required data. For instance, by using the *identify* tool in the navigation bar, the user can obtain detailed information about the "Rn sites " layer as shown in Fig. 11.

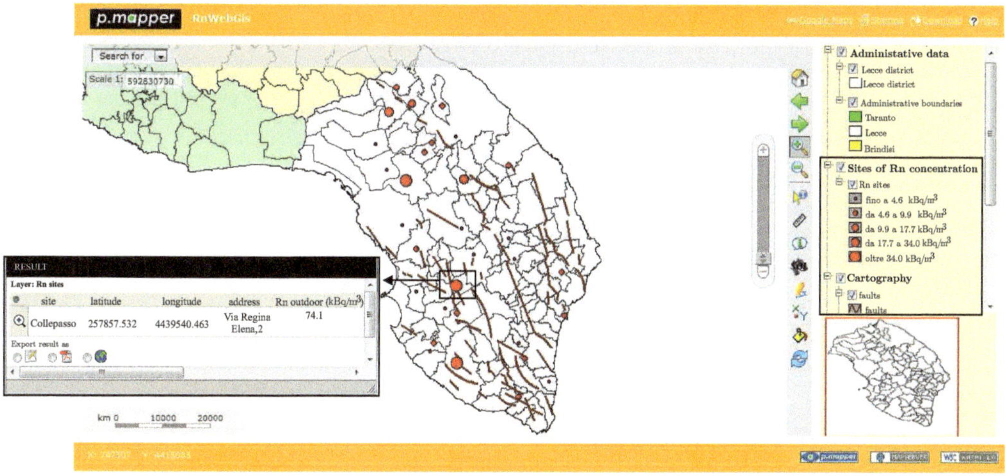

Figure 11. Map Overlay functions: combining layers referred to administrative data, Rn concentrations sites and tectonic faults map. The identify tool highlights an example of Rn site (Collepasso municipality) with high Rn concentrations. Note that the red points indicate the sample locations and their size is proportional to the measured Rn concentration.

Another example of query is obtained by using the *search for* tool, in top left of the *p.mapper* Framework (version 4.1.1.) as shown in Fig. 12.

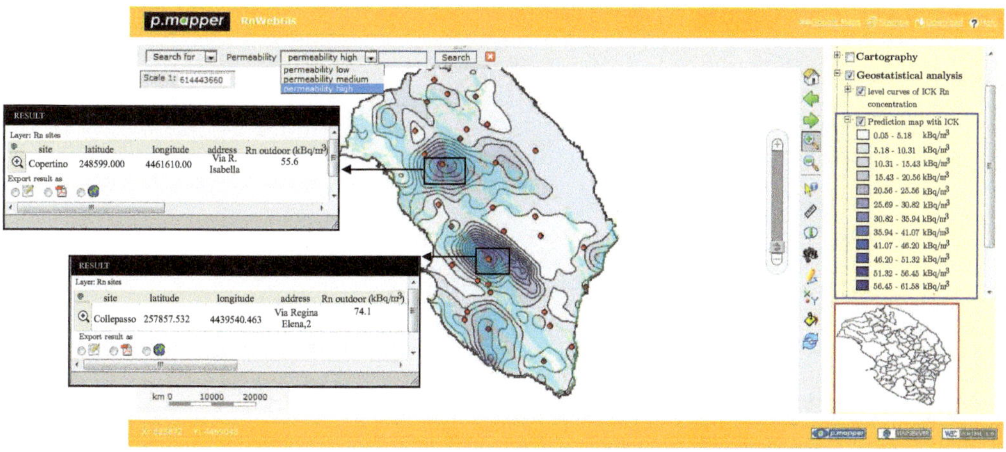

Figure 12. "Search for" tool with the visualization of the layer referred to "high" permeability overlapped on the "Rn site", "level curves of ICK" and "Prediction map of ICK" layers. The identify tool highlights two Rn sites (Collepasso and Copertino municipalities) with high Rn concentrations.

This function has been used to query the database associated with the layer "Permeability" and visualize, by using the thematic map, the study area with high level of permeability overlapped to the layers "level curves of ICK" and "Prediction map of ICK". In this case, the user can identify areas of high level of permeability close to area with high Rn concentrations.

Interactive navigation of maps is possible by using the tools implemented in the RnWebGIS, such as the use of *zooming/panning* available by selecting the object on the map.

On the other hand, in order to ensure the interoperability (i.e. exchange and/or sharing) of geographic data, it has been relevant to build a RnWebGIS according to the specifications of the services, defined by the *Open Geospatial Consortium* (OGC). Indeed, by using the features implemented in the RnWebGIS, the user can also integrate data located on different servers as well as data of different formats. This is possible by recalling the WMS given by the OGC that avoids duplication of data and, at the same time, provides updated geographical data which are certified by shared *standards*.

For example, it is worth highlighting that, by using the appropriate WMS, an orthophoto of Lecce district can be integrated with the layer "Prection map of OCK " as shown in Fig. 13.

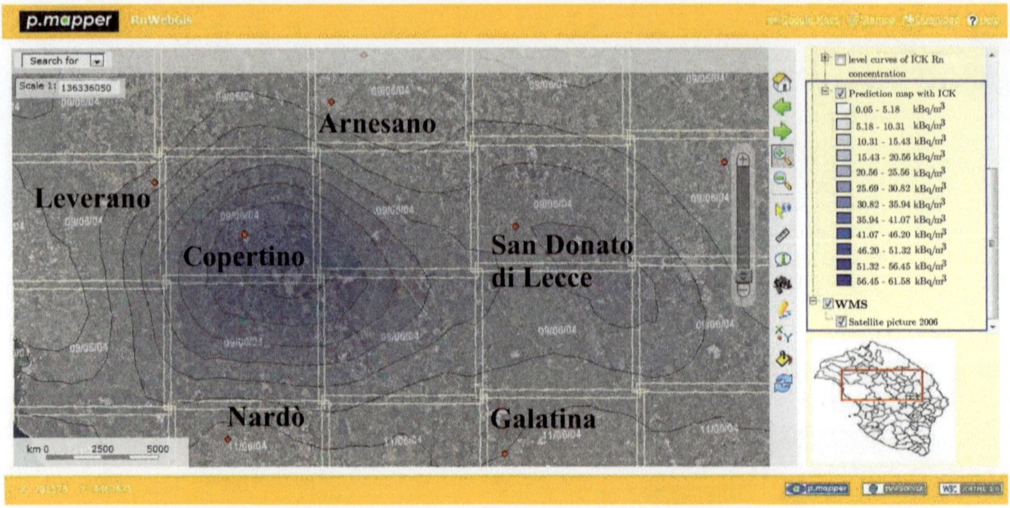

Figure 13. Orthophoto of Lecce district, obtained by the WMS and integrated with the "Rn sites", "level curves of ICK" and "Prediction map of ICK" layers.

6. Conclusions

In this paper, after introducing the usefulness of a GIS supported by geostatistical results, and reviewing some geostatistical modeling and predictions techniques used in the univariate and multivariate cases, a case study concerning a thorough geostatistical analysis of the Rn concentrations in soil gas over Lecce district has been discussed. In particular, four different spatial interpolation techniques, that is Ordinary Kriging, Log-normal Kriging, Cokriging with Indicator variable and Kriging with Varying Local Means, have been used in order to predict the Rn concentration levels over the study area. Then, an *ad hoc* WebGIS, called RnWebGIS, has been proposed. The map of the Rn predictions obtained by using Cokriging with Indicator variable, which performed more accurate predictions than the ones obtained by using the other methods, has been stored into the WebGIS, together with several information about the geo-lithological features (such as permeability and tectonics maps), as well as administrative characteristics (such as demographic data and municipal boundaries) of the study area.

The proposed RnWebGIS highlights the usefulness of the integration between GIS and advanced geostatistical tools, in order to support the management of a sustainable development by taking into account even the risk of exposure to Rn pollution.

Author details

Veronica Distefano, Sandra De Iaco*, Monica Palma, Donato Posa, Alessandra Spennato

*Adress all corresspodence to: s.deiaco@economia.unile.it

University of Salento, Department of Management, Economics, Mathematics and Statistics, Italy

References

[1] Anselin, L. (1992). *Spatial Data Analysis with GIS*, National Center for Geographic Information and Analysis, University of California, USA.

[2] Anselin, L. (2000). Computing Environments for Spatial Data Analysis, *Journal of Geographical Systems* **2**: 201–220.

[3] Bivand, R.S. & Gebhardt, A. (2000). Implementing Functions for Spatial Statistical Analysis using the R Language, *Journal of Geographical Systems* **2**: 307–311.

[4] Beaubien, S.E., Ciotoli, G. & Lombardi, S. (2003). Carbon dioxide and radon gas hazard in the Alban Hills area (Central Italy), *J. Volcanol. Geotherm. Res.* **123**: 63–80.

[5] Beck, H.L. & Gogolak, C.V. (1979). Time-dependent calculations of the vertical distribution of 222Rn and its decay products in the atmosphere, *J. Geophys. Res.* **84**, doi: 10.1029/JC084iC06p03139.

[6] Boots, B. (2000). Using GIS to Promote Spatial Analysis, *Journal of Geographical Systems*, **2**: 17–21.

[7] Bottai, L., De Filippis, T., Mari, R. & Rocchi, L. (2010). *Servizi WebGIS sviluppo sostenibile: la condivisione dell'informazione meteorologica ed ambientale in Toscana*, Proceedings of the 14th National Conference ASITA, Brescia, Italy.

[8] Burrough, P.A. (2001). GIS and Geostatistics: Essential Partners for Spatial Analysis, *Environmental and Ecological Statistics* **8**: 361–377.

[9] Burrough, P.A. & McDonnell, R.A. (1998). *Principles of Geographical Information System*, Oxford University Press.

[10] Chilés, J. & Delfiner, P. (1999). *Geostatistics - Modeling spatial uncertainty*, Wiley.

[11] Cressie, N. (1993). *Statistics for Spatial Data*, Wiley, New York.

[12] De Iaco, S. & Posa, D. (2013). Positive and negative non-separability for space-time covariance models, *J. Stat. Plan. Infer.* **143**: 378–391.

[13] De Iaco, S., Posa, D. & Myers, D.E. (2013). Characteristics of some classes of space-time covariance functions, *J. Stat. Plan. Infer.* **143**: 2002–2015.

[14] De Iaco, S., Distefano, V. & Posa, D. (2013). Geostatistica e GIS, Tecniche e strumenti per il monitoraggio ambientale, in: *Lo sviluppo sostenibile: Ambiente, Risorse, Innovazione, Qualità*, Franco Angeli Editore, Italy.

[15] Delhomme, J.P. (1978). Kriging in the hydrosciences, *Adv. Water Resources* **1**: 251–266.

[16] Durrani, S.A. & Badr, I. (1995). Geostatistically controlled field study of radon levels and the analysis of their spatial variation, *Radiat. Meas.* **25**, 565–572.

[17] Gillavry, E.M. (2000). *Carthographic aspects of WebGIS*, Software Department of Cartography, Utrecht University.

[18] Goodchild, M.F. (1987). A spatial analytical perspective on geographical information systems, *Int. J. Geogr. Inf. Sci.* **1**: 327–334.

[19] Goodchild M.F. (1992). Geographical data modeling, *Comput. Geosci.* **18**: 401–408.

[20] Goodchild, M.F., Haining, R., Wise, S., Arbia, G., Anselin, L., Bossard, E.G., Brunsdon, C., Diggle, P., Flowerdew, R., Green, M., Griffith, D.A., Hepple, L., Krug, T., Martin, R.J. & Openshaw, S. (1992). Integrating GIS and Spatial Data Analysis: Problems and Possibilities, *International Journal of Geographical Information Science* **6**: 407–423.

[21] Goovaerts, P. (1997). *Geostatistics for natural resources evaluation*, Oxford University Press, New York.

[22] Journel, A.G. (1983). Non-parametric estimation of spatial distribution, *Mathematical Geology* **15**: 445–468.

[23] Journel, A.G. & Huijbregts, C.J. (1981). *Mining Geostatistics*, Academic Press, London.

[24] Kropla, B. (2006). *Beginning MapServer: Open Source GIS development*, Berkeley Apress Ed.

[25] Matheron, G. (1962). *Traité de Géostatistique appliquée.* Mémoires du Bureau de Recherches Geologiques et Minieres, Editions Technip, Paris, Tome I, n.14.

[26] Matheron, G. (1965). *Les variables régionalisées et leur estimation: une application de la théorie des fonctions aléatoires aux sciences de la nature*, Masson, Paris.

[27] Matheron, G. (1976). A simple substitute for conditional expectation: the disjunctive kriging. In: *Advanced Geostatistics in the mining industry*, Guarascio M. et al. (eds): 221:236.

[28] Monestiez, P., Courault, D., Allard, D. & Ruget, F. (2001). Spatial interpolation of air temperature using environmental context: Application to a crop model. *Environmental and Ecological Statistics* **8**: 297–309.

[29] O'Dea, J. & Dowdall, M. (1999). Spatial analysis of natural radionuclides in peat overlying a lithological contact in Co. Donegal Ireland, *J. Environ. Radioactiv* **44**: 107–117.

[30] Oliver, M.A. & Khayrat, A.L. (2001). A geostatistical investigation of the spatial variation of radon in soil, *Comput. Geosci.* doi:10.1016/S0098-3004(00)00133-3.

[31] Porstendorfer, J. (1994). Properties and behavior of radon and thoron and their decay products in the air, *J. Aerosol Sci.* **25**: 219–263.

[32] Posa, D. & De Iaco, S. (2009). *Geostatistica. Teoria e Applicazioni*, Giappichelli Ed., Turin, Italy.

[33] Wackernagel, H. (1988). *Geostatistical techniques for interpreting multivariate spatial information*, in: Quantitative Analysis of Mineral and Energy Resources, Chung, C.F. et al. (Eds), 349–409.

[34] Wackernagel, H. (2003). *Multivariate Geostatistics: an introduction with applications*, Springer, Berlin.

[35] Wang, J., Krudy, G., Hou, T., Zhang, W., Holland, G., Xu, X., (2007). *Development of reliable aqueous solubility models and their application in druglike analysis*, Journal of chemical information and modeling. 47(4): 1395-1404.

[36] Wesseling, C.G., Karssenberg, D., Burrough, P.A. & Van Deursen, W.P.A. (1996). Integrating Dynamic Environmental Models in GIS, *Transactions in GIS* **1**: 40–48.

[37] Wise, S., Haining, R. & Ma, J. (2001). Providing Spatial Statistical Data Analysis Functionality for the GIS User: The SAGE Project, *International Journal of Geographical Information Science* **15** (3): 239–254.

[38] Yang, Y., Zhu, J., Tong, X., Wang, D., (2009). The spatial pattern characteristics of soil nutrients at the field scale, Computer and Computing Technologies in Agriculture II **Vol. 1**, Springer US: 125–134.

[39] Zhu, H.C., Charlet, J.M. & Poffijn, A. (2001). Radon risk mapping in southern Belgium: an application of geostatistical and GIS techniques, *Sci. Total Environ.* **272**: 203–210.

2

Indoor Air Quality and Risk Factors Associated with Respiratory Conditions in Nigeria

Godson Rowland Ana, Oyewale Mayowa Morakinyo and Gregory Adekunle Fakunle

Additional information is available at the end of the chapter

1. Introduction

Clean air is a basic requirement for life and healthy living. The quality of air in homes, offices, schools, day care centres, public buildings, health care facilities and other private and public buildings where people spend over 80% (3/5) of their time daily is crucial for healthy living and people's well-being. The National Health and Medical Research Council (NHMRC) defines indoor air as 'air within a building occupied by people of varying states for a period of at least one hour' (NHMRC, 1996). Buildings covered by this definition include homes, schools, restaurants, public buildings, residential institutions, offices, etc. (www.arf.org.au)

Indoor Air Quality (IAQ) is an increasing concern in the world today. In fact, the mere presence of people in a building can significantly alter indoor air quality. Indoor air pollution is present in virtually every indoor space, with the exception of strictly controlled and sterile spaces in pharmaceutical, medical and research facilities. Indoor pollutants may originate from human activities, building materials and carpets; they may also penetrate from outdoor environments by forced ventilation, diffusion or infiltration [8]. Humans are inevitably exposed to these pollutants, considering the amount of time spent indoors, but the influence of the pollution on human health may vary, depending on age, sex, nutritional status, physiological conditions, and individual predisposition.

There is increasing evidence linking indoor air pollution to increased risk of respiratory tract infections, exacerbations of inflammatory lung conditions, development of chronic obstructive lung disease, cardiac events, stroke, eye disease, tuberculosis, cancer and hospital admissions especially in women and children who are the most exposed [40, 54, 59, 63].Acute Respiratory Infections (ARIs) are responsible for one-third of hospital admissions and 20%–30% of deaths among children under 5 years [19, 48] in developing countries.

However, information regarding IAQ in Nigeria is scanty. Aside from the data collected by a few individuals and corporate organisations at scattered locations, there is no comprehensive and empirical database on the magnitude of the health risks and deleterious effects associated with exposure to poor IAQ by occupants in different built environments and the general population in Nigeria.

2. Brief profile of Nigeria

Nigeria is located in western Africa on the Gulf of Guinea and has a total area of 923,768 km^2. It shares a 4,047 km border with Benin (773 km), Niger (1497 km), Chad (87 km) and Cameroon (1690 km), with a coastline of at least 853 km. Nigeria lies between latitudes 4° and 14°N, and longitudes 2° and 15°E. Nigeria has about174 million inhabitants and is the most populous country in Africa and the seventh most populous country in the world. The country has over 500 ethnic groups, of which the three largest are the Hausa, Yoruba and Igbo. Nigeria is the 12th largest producer of petroleum in the world and the 8th largest exporter, and has the 10th largest proven reserves. Petroleum plays a large role in the Nigerian economy, accounting for 40% of GDP and 80% of Government earnings.

3. Indoor air pollution

Indoor air pollution refers to the amount of chemical, biological and physical contaminants in the air inside a building. There is a wide range of indoor air pollutant sources in houses including building materials, furniture, central heating and cooling systems, household cleaning products, paints, tobacco smoke, office machines and a variety of other products used in daily activities (EPA, 2009).

Nigeria, unlike the developed countries is faced with several environmental, social and economic challenges such as inadequate electric power supply, poor waste disposal system, air pollution, water pollution, noise pollution, unemployment, inadequate of water supply, etc. [67]. Most households in Nigerian cities operate small capacity fossil fuel electric power generators for electricity supply (ECN, 2009).

A study carried out by [67], revealed that small household generators in Nigeria operate an average of 6 h daily, while the average distance of generator away from building was 5.6 m. These alongside poor ventilation have influenced the quality of indoor air in the households [46].

Another major source of pollution in the indoor environment is household combustion of coal or biomass for cooking and heating. It is estimated that more than 50% of the world's population depends on animal dung, wood, crop waste or coal to meet their most basic energy needs (WHO, 2005). Firewood combustion for cooking is a common practice in most rural communities of developing nations including Nigeria. Its use as an energy source is widely reported to impinge on the environment, particularly the quality of air and the health of populations especially women [5].

Based on a study designed to determine exposure to emissions from firewood cooking stove and the pulmonary health of women in Olorunda community in Ibadan, Nigeria, [5] found that the mean CO levels for fire wood stove emissions at six kitchen locations combined were 20.0±0.8 ppm (morning) and 24.2 ± 0.5 ppm (evening). These values were 2 and 2.4-folds higher, respectively, than the World Health Organization (WHO) limit of 10 ppm. The mean SO_2 levels for firewood stove emissions at the six locations were 0.505 ppm (morning) and 0.683 ppm (evening). These values were 6.3- and 8.5-folds higher, respectively, than the WHO limit of 0.08 ppm. The mean NO_2 levels for firewood stove emissions at the six locations were 0.517 ppm (morning) and 0.81 7 ppm (evening). These values were 25- and 41-fold higher, respectively, than the WHO limit of 0.02 ppm(Fig 1).

In this study, although the levels of NO_2 were low, there was a significant negative correlation between NO_2 and lung function (r = − 0.488), which indicated that chronic exposure to NO_2 is likely to impair lung functions (Ana *et al.*, 2012).

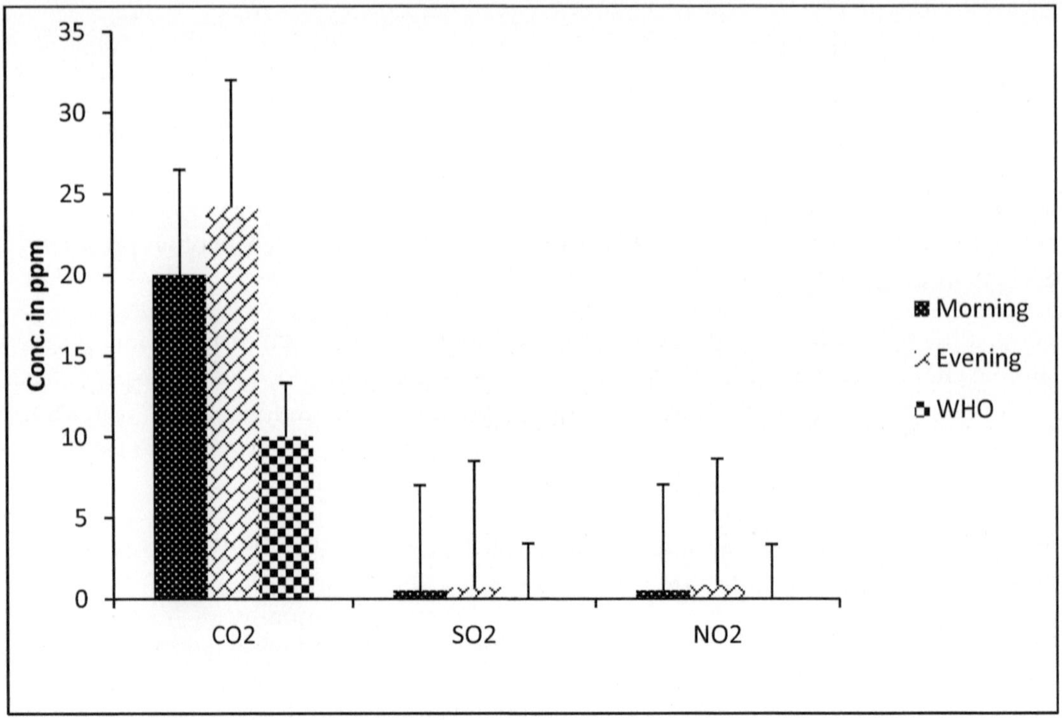

Source : Ana *et al.*, 2012

Figure 1. Mean morning and evening gaseous concentrations at six kitchen locations in comparison with WHO limits

4. Housing and indoor air quality

Housing is a place people spend most of their time and one of 'the main settings that affect human health' [12]. It is well established that housing is a 'key determinant of health' [62]. Good housing and good health go together [62]. Housing, as a neglected site for public health action, has been identified in a number of recent publications (Breysse *et al.*, 2005). Housing, however, encompasses a large range of factors, including biological (mould, cockroaches, dust mites, etc.), chemical (tobacco smoke, paints, etc.) and structural (moisture, heat ventilation, etc.).

A large number of the population in Nigeria resides in houses with poor housing conditions (as shown in Figure 2.). Provision of adequate housing has been a problem in developing countries like Nigeria. Shortage of decent accommodation, particularly in cities, has resulted in individuals occupying bungalows or rooms above specified capacity. More so, in the quest to provide themselves with urgent accommodation, people have resorted to building houses outside required specifications and/or at indecent locations. Because of this problem, there is no available information on the indoor air quality of living homes in the country.

5. Indoor air environment

Indoor environments are fundamental environmental factors capable of impacting health [16, 32]. Air quality of indoor environments is one of the main factors affecting the health, well-being and productivity of people. The effect on health rises as exposure to and density of air pollution increases [35, 42].

In indoor air environments (non-industrial), the most important source of airborne bacteria is the presence of humans (Kim and Kim, 2007). Major indoor activities such as talking, sneezing, coughing, etc. can concentrate airborne biological particulate matter in the indoor environment. House dust, pets and their bedding material, wood materials and furniture stuffing could serve as reservoir from which spores of *Alternaria, Aspergillus, Botrytis, Cladosporium, Penicillium, Scopulariopsis* spp. and yeast cells are occasionally released into the indoor air [17].

6. Bioaerosols in the indoor air environment

Bioaerosols consist of aerosols containing microorganisms (bacteria, fungi, viruses) or organic compounds derived from microorganisms (endotoxins, metabolites, toxins and other microbial fragments) (Heikkienenet *al.*, 2005). Aerosols of biological origin form a significant portion of the total atmospheric aerosols, sometimes reaching up to 50% of all aerosol particles [37].

Bioaerosols vary in size (20 nm to >100 nm) and composition depending on the source, aerosolisation mechanisms and environmental conditions prevailing at the site [53]. The inhalable fraction ($PM_{2.5}$) is of primary concern because it is the most susceptible portion of the bioaerosols to reach the deeper parts of the respiratory system (Pope *et al.*, 1995).

(a) (b)

(c) (d)

Plate 1. Typical housing structure in urban (a and b) and rural (c and d) settings in Nigeria

Indoor air contains a complex mixture of bioaerosols such as fungi, bacteria and allergens along with non-biological particles (e.g., dust, smoke, particles generated by cooking, organic and inorganic gases) [33]. Airborne microorganisms might pose an environmental hazard when present in high concentrations in indoor environments resulting in health problems [68].

Bioaersosols contribute to about 5%--34% of indoor air pollution [66]. The source of bioaerosols in indoor air includes furnishing and building materials, microbiological contamination within the walls and ceilings and floor activities. The outdoor environments are oftentimes major sources of indoor bioaerosols as particles are transferred indoor through openings of the building envelope (windows, doors). However, one of the most important factors affecting indoor air quality is how the building is heated, ventilated, air-conditioned and its occupancy [9].

6.1. Bacteria in indoor air environment

Airborne bacteria are major issues in indoor environments. The potential health risks resulting from exposure to airborne bacteria can occur at workplaces and residential locations [28]. A study conducted in Nigeria by [7] in day care centres, reported that the genera of bacteria isolated from the indoor environment included *Pseudomonas, Staphylococcus* and *Bacillus* spp., in both wet and dry seasons.

Elsewhere, the most important bacterial strains found in an indoor atmosphere are representatives of the genera *Bacillus, Micrococcus, Kocuria and Staphylococcus, Streptomyces albus, Pantoea agglomerans, Pseudomonas chlororaphis, Arthrobacter globiformis, Thermoactinomyces vulgaris and Corynebacterium* spp.[23].

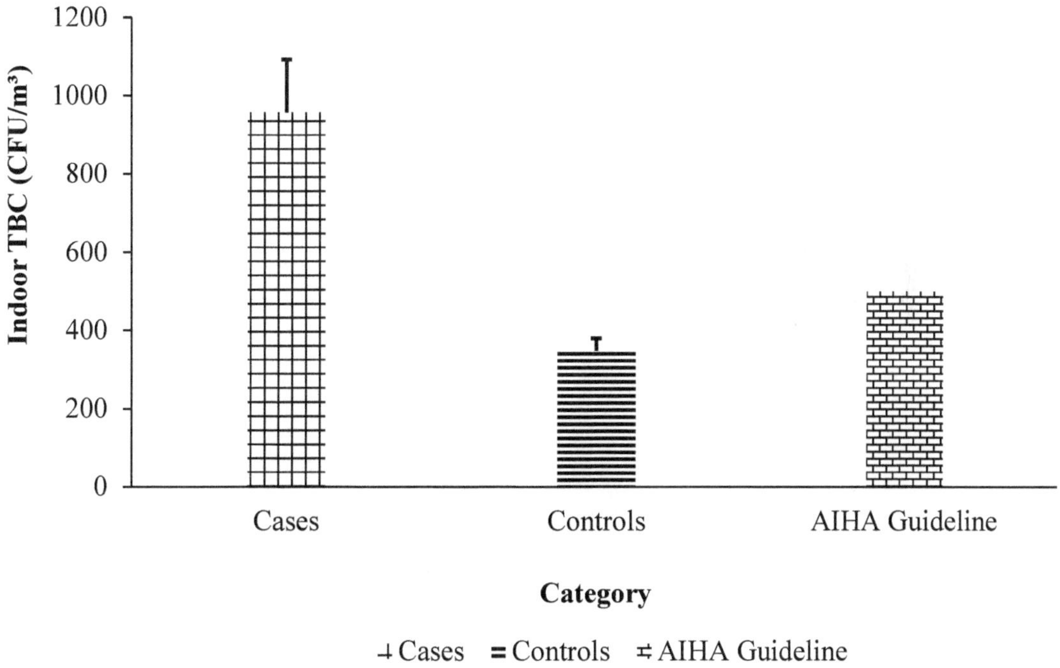

Source: Ana et al., 2013

Figure 2. Mean Indoor TBC among cases and controls as compared with AIHA Guideline

Also, in a study designed to determine the burden of airborne microbes in indoor environment that predispose children under the age of 5 to acute respiratory infections, the indoor airborne bacterial load among cases ($9.6x10^2 cfu/m^3$) was higher than the acceptable limit

(≤5.0x10^{2}cfu/m^3) proposed by the American Industrial Hygiene Association (AIHA, 2001) for residential locations compared to controls (3.5x10^{2}cfu/m^3) [7] (Fig. 2).

Like fungi, bacterial growth takes place in the presence of sufficient moisture. In the indoor environments, sources of required moisture for microbial growth include drip pans of air-cooling devices, freezers or refrigerator, humidifiers and wet surfaces of bathrooms and kitchen [45]. Water may also originate from leaks in plumbing or roof, or be a result of condensation or harsh rain or snowing [65]. Standing water is a potential source of bacteria, and once disturbed, bacteria may become airborne [65].

6.2. Fungi in indoor air environments

The building-associated fungi consist of filamentous microfungi (moulds) and yeasts. The most common fungal genera occurring in indoor environments are *Penicillium, Aspergillus, Cladosporium, Alternaria* and yeasts [36].

In the same study carried out by [6] they showed that genera of fungi isolated in indoor environment of day care centres included *Aspergillus, Penicillum, Geotricum, Fusarium* and *Candida* spp., and in the dry season *Aspergillus, Penicillum, Geotricum* and *Fusarium* spp.,

Their presence in indoor air is as a result of transportation from outside environment via building materials, carpets, furniture, wallpapers, etc. Ventilation and air-conditioning systems are another common way of penetraton of fungi into the buildings. The rate of further growth, spreading and multiplication depends exclusively on moisture content in indoor air, regardless of the type of surface [30].

Natural food source for fungi vary from plant, animal and human particles in house dust, to fragments of construction materials such as floor and wall textile coverings, furniture, residue of cooking traces, food storage and paper materials. Since these materials occur in good quantity in every building, and considering that optimal temperature for fungi growth ranges from 10–35 ºC, the only limiting factor for development of fungi and mould contamination is dampness [31].

Fungi may be extremely harmful for human health, but may also destroy the building itself (Figure 4.), particularly wooden parts, such as roofs and other materials. Some fungi species produce strong allergens, which initiate immune reaction type I (IgE mediated). For example, the indoor contamination with *Alternaria, Penicillium, Aspergillus* and *Cladosporium* spp., is related to asthma and other allergic respiratory diseases. Some of these species, such as *Penicillium* and *Aspergillus* can also induce type III allergy (IgG mediated), while at high concentrations, may also initiate combined type III and IV reaction manifested as hypersensitivity pneumonitis. Major fungal allergens are isolated and identified (such as Cla h I from *Cladosporium herbarum,* Alt a I and Alt a II from *Alternaria alternata* and Asp f I and Asp f III from *Aspergillus fumigatus*) [22].

Fungi mycotoxins have strong genotoxic, carcinogenic and immunotoxic potential. The carcinogenic effects of aflatoxin (mycotoxin produced by *Aspergillus flavus* and *Aspergillus parasiticus*) are well known. The most important mycotoxins related to indoor air contamination are trichothecenes, generated by fungi *Stachybotrys chartarum* (macrocyclic trichothecenes, trichodermin, sterigmatocystin and satratoxin G) [10].

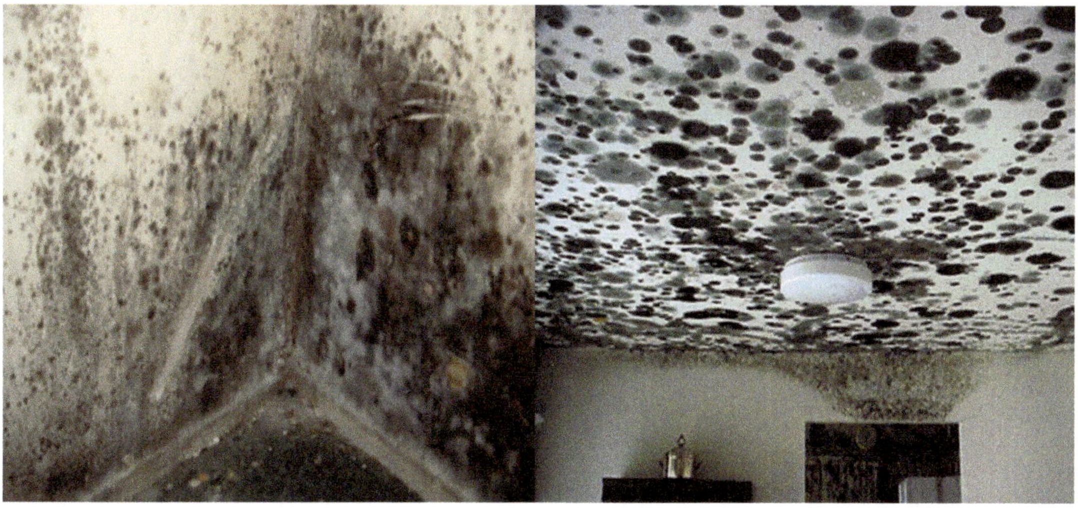

Plate 2. Fungi on the wall and ceiling of a typical indoor environment

6.3. Viruses in indoor environment

Viruses are small (20-400 nm), obligate intracellular parasites. They represent a common cause of infectious disease acquired indoors, as they are easily transmitted especially in crowded, poorly ventilated environments [72].

Airborne transmission of viral agents occurs when viruses travel on relatively large respiratory droplets (> 10 μm) that people sneeze, cough or exhale during conversation or breathing (primary aerosolisation). A single cough can release hundreds of droplets while a single sneeze can release about 40 000 droplets at speeds of up to 50--200 miles per hour with each droplet containing millions of viral particles. Aerosol droplets travel only short distances (1-2 meters) before settlings on surfaces, where viruses can remain infectious for hours or days.

Poorly ventilated and crowded indoor environments are favourable conditions for viral transmission. Hospitals, in particular, are environments where viral aerosols can be particularly hazardous, since patients tend to be especially prone to infection due to pre-existing illness. Elderly patients, children, cancer patients, immunocompromised or immunosuppressed patients are most at risk.

6.4. Protozoa in indoor environment

Protozoa may also be present in indoor air in damp buildings. [77] detected amoebae in 22% of the 124 samples of various materials collected from buildings with evident moisture damage. Field studies on the presence and concentrations of protozoa in indoor air, as well as health aspects of these microorganisms in given conditions, are still lacking, with the exception of one in vitro study conducted by the same authors, who suggested that amoebae act synergistically with certain bacteria, enhancing their cytotoxic and pro-inflammatory potential [78].

7. Prevalence of respiratory problems in Nigeria

Acute respiratory infections are a major cause of morbidity and mortality among Nigerian children. It was estimated that pneumonia accounted for 20% of deaths in children under the age of 5 years in Nigeria between the year 2000 and 2003 (WHO, 2007).

A prospective cohort study in Ilorin, Nigeria, reported that the rate of acute respiratory infection was three episodes per child per year with pneumonia being responsible for 1.3 episodes per child per year [27]. In another hospital-based study in Ibadan, 28.4% of children admitted to the hospital with acute lower respiratory tract infection had acute bronchiolitis with respiratory syncytial virus being the most common viral aetiologic agent [39].

There is scanty data on the bacterial aetiology of pneumonia in Nigerian children.According to WHO, (2002), about 20% of all deaths in children under 5 years are due to Acute Lower Respiratory Infections (ALRIs - pneumonia, bronchiolitis and bronchitis); 90% of these deaths are due to pneumonia. There is a seasonal variation in acute respiratory infections in Nigerian children with more episodes occurring during the rainy season [5, 27, 39].

8. Indoor environmental risk factors for respiratory conditions

8.1. Overcrowding

Many children are exposed to very crowded conditions at home, schools (Figure 5.) etc., and this increases risk of transmission of illness. Most studies in developing countries have found that the average area of habitable space per person is well below the WHO recommendation of 12m^2 [15].

According to a study by [7], the mean number of occupancy among children under the age of 5 admitted in a tertiary health facility for ARIs was 6.0±1.5 as compared to 4.0±1.0 among controls. A positive association was found between the level of occupancy and indoor total bacterial count. A similar study by [70] recorded the highest bacterial burden in an over-crowded environment. This suggests that the number of persons in the household is directly proportional to the level of bacteria build-up in the indoor environment.

8.2. Environmental tobacco smoke among parents

More than 150 published studies have shown a significant relationship between Environmental Tobacco Smoke (ETS) and respiratory illness in children. Meta-analyses revealed a strong evidence for associations between both prenatal maternal smoking and postnatal ETS exposure and risk of ARI in children [21].

[6] in their study indicated that parental smoking or any other smoker in the house as a risk factor for ARIs in children under the age of 5 (OR = 4.7; CI = 0.9-2.17, $p<0.05$). This could be due to the accumulation of emissions from cigarette smoking in the indoor environment as a result of inadequate ventilation [13].

8.3. Household biomass fuel

Biomass fuel is any material derived from plants or animals which is deliberately burnt by humans. Firewood combustion for cooking is a common practice in most rural communities of developing nations including Nigeria [4]. Majority of the households burn biomass fuels in open fireplaces, consisting of simple arrangements as three rocks (Figure 7.), a U-shaped hole in a block of clay, a pit in the ground or in poorly functioning earth or metal stoves. The process of combustion in most of these stoves is incomplete, resulting in substantial emissions which, in the presence of poor ventilation, produce very high levels of indoor pollution (WHO, 2000).

Plate 3. Overcrowding condition in a typical school setting

Many of the substances in biomass smoke have adverse health effects. The most important ones are carbon monoxide, nitrous oxides, sulphur oxides, formaldehyde and polycyclic organic matter, including carcinogens such as benzo[a]pyrene. Particles with diameters <10 microns (PM_{10}), and particularly those <2.5 microns in diameter ($PM_{2.5}$), can penetrate deeply into the lungs (USEPA, 1997).

A Nigerian study on the concentration of indoor air pollutants due to the use of firewood for cooking and its effects on the lung function of women living in the selected homes found that the concentration of PM_{10} and gaseous emissions such as CO and NO_2 significantly exceeded the WHO limits by several folds. Chronic exposure to high level of the indoor air pollutants particularly the respirable particulate matter could possibly compromise the lung function status of women [4].

[47] in their study of the effects of stove intervention on household air pollution and the respiratory health of women and children in rural Nigeria, demonstrated an association between cooking with biomass fuel, increased prevalence of respiratory symptoms and presence of obstructive lung disease. The high frequency of respirator y and other exposure-related symptoms seen in mothers and children at baseline (pre-intervention) were substantially reduced a year later following replacement of the traditional stoves with low-emission stoves (Fig.3). This observation is consistent with the report of higher frequency of respiratory symptoms, especially cough in women exposed to biomass smoke in Mozambique [24], Nepal [60], Pakistan [61], Mexico [56] and Guatemala [20, 64].

Although the incidence of some respiratory symptoms may be lower in children compared to mothers, it is evident that the children are also exposed to the same $PM_{2.5}$ and CO levels, as 94.1% of the mothers always have their children with them during cooking [47]. Studies have also suggested that children exposed to household air pollution early in life may have impaired lung development, indicating that the impact of exposure to biomass smoke may continue into adulthood [2].

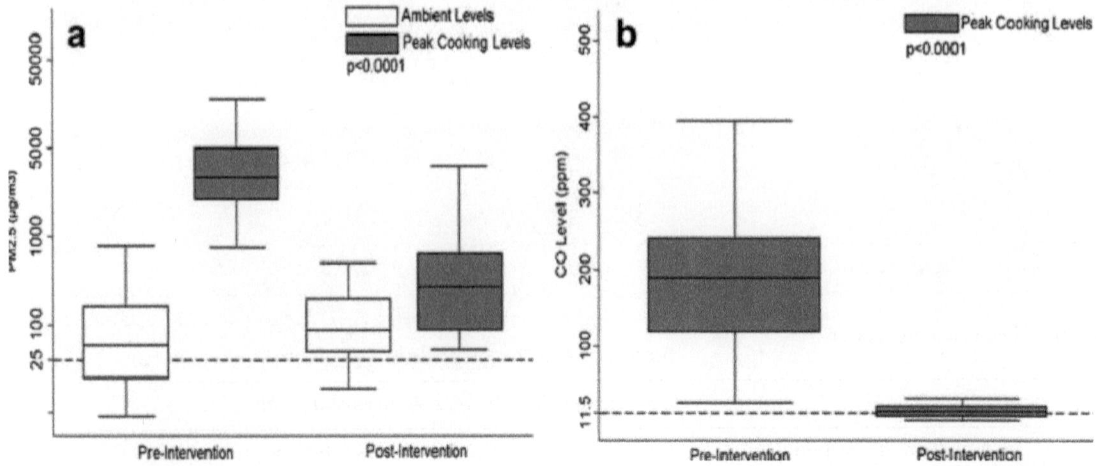

Source: Oluwole et al., 2013

Figure 3. Comparison of household particulate matter (a) and carbon monoxide (b) concentration levels before and 1 year after distribution and monitored use of low-emission stoves. The dotted lines represent WHO recommended levels. It has been highlighted to show that cooking time concentrations of $PM_{2.5}$ and CO in all households sampled were significantly above the recommended levels before distribution and monitored use of low-emission cooking stoves

Over 40 % of the participating mothers and children in our study had mild to moderate obstructive defects. This impaired lung function is likely caused by particulate matter (PM) in firewood smoke, which is believed to induce oxidative injury through its ability to carry adherent metals into the lungs and cause inflammation [57].

Source: Oluwole et al., 2013

Plate 4. Simple arrangement of three stones where biomass burns in an indoor environment

Exposure to these pollutants is reported to be higher in women and children (Figure 8.) [1, 11]. It has been shown that exposure to biomass fuel smoke is responsible for a number of respiratory diseases such as Acute Respiratory Infections (ARI), Chronic Obstructive Pulmonary Disease (COPD), Tuberculosis and Asthma; Low Birth Weight; Cataract and Blindness [14]. The adverse effects on respiratory health of products of incomplete solid-fuel combustion are summarized in Table 1.

Compound	Potential Health Effect
Breathable particulate matter	Wheezing, exacerbation of asthma, respiratory infections, chronic bronchitis and chronic obstructive pulmonary disease.
Carbon monoxide	Low birth weight, increase in perinatal deaths.
Polycyclic aromatic hydrocarbons	Lung cancer, cancer of mouth, nasopharynx and larynx.
Nitrogen oxides	Wheezing, exacerbation of asthma, respiratory infections, reduced lung function in children.
Sulphur oxides	Wheezing, exacerbation of asthma, exacerbation of chronic obstructive pulmonary disease, cardiovascular disease.

Source: Perez et al., 2011

Table 1. Adverse effects on respiratory health associated with indoor air pollution

Source: Oluwole et al., 2013

Plate 5. Mother and child exposed to biomass emissions in an indoor environment

8.4. Dampness in indoor environments

Numerous studies have examined the potential association between damp housing conditions and respiratory ailments among occupants. It was discovered that increase humidity leads to increase mould growth and exposure, which could result in asthma and other respiratory conditions. The presence of home dampness has been reported to affect about 38% of Canadian homes [18]. A Canadian study found that children living in damp or mouldy homes were 32% more likely to have bronchitis [18] (secure.cihi.ca). Substantial problems have been identified in some First Nations communities due to a combination of inappropriate housing design, poor construction, inadequate maintenance and poor ventilation [41].

8.5. House dust mites and cockroaches in indoor environments

House Dust Mites (HDMs) thrive in the dust of homes, particularly in the presence of high indoor relative humidity (secure.cihi.ca). In order to survive and multiply they require a relative humidity in excess of 45%–50%, but their activity, including feeding and maturation is more rapid at higher rates of relative humidity, which was confirmed in field studies [79].

Cockroaches are implicated as a major cause of asthma among inner city children in the U.S., resulting in increased hospital admissions, school absenteeism, and unscheduled medical visits for asthma [58].

8.6. Sanitation and housing quality

[15] found that children with respiratory illness come from houses with poorer sanitation than controls, while in developed countries promotion of hand washing has been associated with reduced incidence of respiratory illness.

8.7. Indoor air meteorological conditions

A case-control study carried out in Ibadan among houses of children under the age of 5 revealed that a higher proportion of houses visited among cases and controls recorded indoor relative humidity (RH) values above the comfort level (30%–60%)(Fig. 4). The high RH (above comfort level) observed among a large proportion of houses among cases could be as a result of high moisture content. The mean indoor and outdoor air RH among cases (69.6 ± 4.7; 67.2 ± 5.0) was found to be higher than among controls (63.1 ± 6.5; 66.1 ± 7.1) [7]. With such high relative humidity levels, microorganisms such as fungi and bacteria can survive on non-living materials including dusts (Choa *et al.*, 2002). High relative humidity above 70% also tends to favour the survival of viruses that infect the membrane of the respiratory tract.

8.8. Inhalable particulate matter burden

Particulate matter (PM) has been recognized in recent years as the most dangerous and widely spread air pollutant [43] affecting more people than any other pollutant and contributing to the risk of developing cardiovascular and respiratory diseases, as well as lung cancer. PM_{10} is associated with increased mortality and morbidity in many cities worldwide and the risk is greatly pronounced among susceptible populations, including the elderly, children and people with pre-existing health conditions [55].

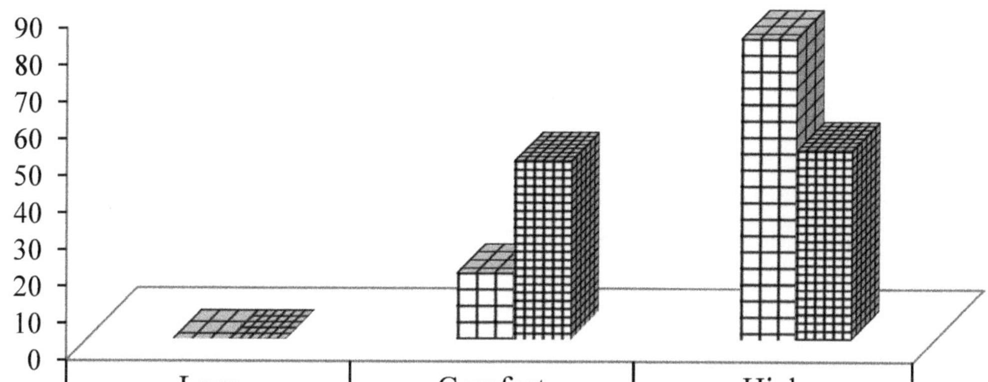

	Low	Comfort	High
Case	0	18.2	81.8
Control	0	48.5	51.5

Key: Low = <30%, Comfort = 30 – 60%, High = >60%
Source: Ana et al, 2013

Figure 4. Indoor relative humidity for houses among cases and controls

[6] in their study of inhalable particulate matter burden in selected day care centres in Ibadan, Nigeria, observed that the mean indoor PM_{10} readings for wet season (73.4 ± 54.4μg/m³) and dry season (296.3 ± 61.6μg/m³) significantly exceeded the WHO guideline limit of 50μg/m³ (Table 2).The indoor/outdoor PM_{10} ratio was 1.38 ± 0.97 and being greater than 1 indicated a possible indoor source of pollution. High particulate matter concentration observed was indicative of the poor indoor air quality condition in the day care centres.

Parameter	Season	Mean ± Standard deviation	p-value
Indoor condition			
Particulate Matter (PM_{10})	Wet	88.6 ± 55.6	<0.001
	Dry	299.8 ± 62.9	
Source: Ana and Umar, 2013			

Table 2. Indoor particulate matter concentration in the wet and dry seasons

9. Conclusion

Everyone has the right to a good standard of living, adequate for the health and well-being of his family. Over the years, man has built houses to protect himself from environmental hazards. Buildings do not always protect their occupants from pollution. The quality of housing conditions plays a decisive role in the health status of the residents. The bioaerosols, dust and gaseous pollutants trapped or growing on the inside may well exceed those outdoors and therefore exacerbate health problems.

A polluted indoor air is a risk factor for human health globally. The magnitude of the health risks and risks factors associated with exposure to poor indoor air quality especially among children under the age of 5 are evident in Nigeria. Overcrowding, the use of generators and biomass fuel have been identified as the critical environmental problems which increases the likelihood of respiratory symptoms and/or obstruction in those exposed on a regular basis in Nigeria.

Health policy changes regarding construction of homes with better ventilated kitchens and use of environmentally friendly, low-emission and energy-efficient cooking stoves, aimed at mitigating the deleterious effects of exposure to indoor air pollutants in mothers and children will significantly improve indoor air quality and reduce the risk of respiratory symptoms.

Indoor Air Quality will continue to be an essential field in environmental health studies and interventions because exposure to varied indoor substances will likely increase in the coming years. Research on the molecular characterisation of indoor bioaerosols, genetic susceptibility to indoor pollutants and their carcinogenic effect are required.

Author details

Godson Rowland Ana[1,2*], Oyewale Mayowa Morakinyo[1] and Gregory Adekunle Fakunle[1,3]

*Address all correspondence to: anagrow@yahoo.com

1 Department of Environmental Health Sciences, Faculty of Public Health, University of Ibadan, Nigeria

2 Climate and Society Programme Centre for Sustainable Development, University of Ibadan, Nigeria

3 Department of Medicine, University College Hospital, Ibadan, Nigeria

References

[1] Albalak, R., Frisancho, A.R. and Keeler, G.J. (1999). Domestic biomass fuel combustion and chronic bronchitis in two rural Bolivian villages. *Thorax*; 54 (11): 1004-1008.

[2] Almond, D. (2006). Is the 1918 influenza pandemic over? Long-term effects of in-utero influenza exposure in the post 1940 USA population? *Journal of Political Economy;* 114(4):672 -712

[3] American Industrial Hygiene Association (AIHA) (2001). Report of microbial growth task force. From http://www.hc-sc.gc.ca/ewh-semt/pubs/air/fungal-fongique/references-eng.php (retrieved 23 January 2009). Fairfax, VA1.

[4] Ana, M.G., Dairo, D., Adebowale, S. and Ana, G. (2012). A hospital-based pilot survey of acute infections among under-Five Children in Ibadan, Nigeria. Proceedings of the Ibadan Sustainable Development Summit; 185 – 191.

[5] Ana, G., Adeniji, B., Ige, O., Oluwole, O., Olopade, C.O. (2012). Exposure to emissions from firewood cooking stove and the pulmonary health of women in Olrunda community, Ibadan, Nigeria. Air QualAtmos Health (online version). doi:10.1007/s11869-012-0183-6

[6] Ana, G.R., Fakunle, G.A. and Ogunjobi, A.A. (2013). Indoor airborne microbial burden and risk of acute respiratory infections among children under five in Ibadan, Nigeria. *Indoor and Built Environment*; 0(0): 1-7.

[7] Ana, G.R.E.E. and Umar, Z.O. (2013). Indoor Air Quality and Respiratory Symptoms experienced among under-five children in Daycare centres in Ibadan North Local Government Area, Oyo State, Nigeria. An unpublished Master's dissertation in the Department of Environmental Health Sciences, University of Ibadan, Nigeria.

[8] Anyanwu, E.C. (2011). Advances in Environmental Health Effects of Toxigenic Mold and Mycotoxins. New York: Nova Science Publishers, Inc.

[9] Bartlett, K.H., Kennedy, S.M., Brauer, M., Van Netten, C. and Dill, B. (2004). Evaluation and determinants of airborne bacterial concentrations in school classrooms. *J Occup Environ Hyg 2004*; 1: 639-647.

[10] Bloom, E., Bal, K., Nyman, E., Must, A. and Larsson, L. (2007). Mass spectrometry-based strategy for direct detection and quantification of some mycotoxins produced by *Stachybotrys* and *Aspergillus spp.* in indoor environments. *Appl Environ Microbiol*; 73(13): 4211-4217.

[11] Boadi, K.O and Kuitunen, M. (2006). Factors affecting the choice of cooking fuel, cooking place and respiratory health in the Accra metropolitan area, Ghana. *Journal of biosocial Science*;38: 403-412.

[12] Bonnefoy, X., Braubach, M., Krapavickaite, D., Ormandy, D. and Zurlyte, I. (2003). Housing conditions and self-reported health status: a study in panel block buildings in three cities of Eastern Europe, *Journal of Housing and the Built Environment*; 18: 329-352.

[13] Brims, F. and Chauhan, A.J. (2005). Air quality, tobacco smoke, urban crowding and daycare: modern menaces and their effects on health. *Journal of Pediatrics Infectious Disease*; 11:152-156.

[14] Bruce, N., Perez-Padilla, R. and Albalak, R. (2000). Indoor air pollution in developing countries: a major environmental and public health challenge. *Bull WHO*; 78: 1080-1092.

[15] Cardoso, M.R., Cousens, S.N., De Goes Siqueira, L.F., Alves, F.M. and D'Angelo, L.A. (2004). Crowding: risk factor or protective factor for lower respiratory disease in developing countries. *Biomedical Journal of Public Health*;3(4):19.

[16] Cobanglu, N. and Kiper, N. (2006). Binaicisolunanhavadatehlikeler. CocukSagligive-HastaliklariDergisi; 49: 71-75.

[17] Cox, C.S. and Wathes, C.M. (1995). Bioaerosols handbook. New York. Retrieved September 12, 2011, from www.alibris.com/search/books/./ Bioaerosols%20Handbook.

[18] Dales, R.E., Zwanenburg, H., Burnett, R. and Franklin, C.A. (1991). Respiratory health effects of home dampness and molds among Canadian children. *American Journal of Epidemiology*;134.2: 196-200.

[19] Demers, A.M., Morency, P., Mberyo-Yaah, F., Jaffer, S., Blais, C. and Somse, P. (2000). Risk factor for mortality among children hospitalized because of acute respiratory infections in Bangui, Central African Republic. *J Pediatr Infect Dis.*; 19: 424-432.

[20] Diaz, E., Bruce, N., Pope, D., Lie, RT., Diaz, A., Arana, B. *et al.* (2007). Lung function and symptoms among indigenous Mayan women exposed to high levels of indoor air pollution. *Int J Tuberc Lung Dis.;* 11(1 2): 1372-1379.

[21] DiFranza, J.R., Aligne, C.A. and Weitzman, M. (2004). Prenatal and postnatal environmental tobacco smoke exposure in children's. *Pediatrics;* 113(4): 1007-1015.

[22] Douwes J. (2005). (1: >3)-Beta-D-glucans and respiratory health: a re-view of the scientific evidence. *Indoor Air;* 15(3): 160-169.

[23] Dutkiewicz, J., Krysinska-Traczyk, E., Skorska, C. *et al* (2002). Exposure to airborne microorganisms and endotoxin in a potato processing plant. *Ann Agric EnvironMed;* 9: 225-235.

[24] Ellegard, A. (1996). Cooking fuel smoke and respiratory symptoms among women in low income areas in Maputo. *Environ Health Perspect;* 104(9):980-985

[25] Energy Commission of Nigeria (ECN) (2009). 60 Nigerians Now Own Power Generators. Adopted from Vanguard Newspaper, 26th January, 2009. [Online] Available at www.energy.gov.ng.

[26] Environmental Protection Agency (EPA) (2009). Residential Air Cleaners. Indoor Air Quality, Available from http://www.epa.gov/iaq/pdfs/residential_air_cleaners.pdf.

[27] Fabule, D., Parakoyi, D.B. and Spiegel, R (1994). Acute respiratory infections in Nigerian children: prospective cohort study of incidence and case management. *J Trop Pediatr;* 40: 279-284.

[28] Halide, A., Ahmet, A., Muserref, T.O. (2009). Indoor and outdoor airborne bacteria in child day-care centers in Edirne City (Turkey), seasonal distribution and influence of meteorological factors", Environmental Monitoring and Assessment; 164: 53-66

[29] Johnson, A.W. and Aderele, W. I. (1996). The association of household pollutants and socio-economic risk factors with the short-term outcome of acute lower respiratory infections in hospitalized pre-school Nigerian children. *Annal Tropical Paediatric;* 12(4): 421-432.

[30] Grant, C., Hunter, C.A., Flannigan, B. and Bravery, A.F. (1989).The moisture requirements of moulds isolated from domestic dwellings. *IntBiodeterior;* 25(4): 259-284.

[31] Green, B.J., Tovey, E.R., Serkombe, J.K., Blachere, F.M., Beezhold, D.H., Schmechel, D. et al. (2006). Airborne fungal fragments and allergenicity. *Med Mycol;* 44(1): 245-255.

[32] Gocgeldi, E., Berdan, M.E., Ucar, M., Turker, T., Istanbulluoglu, H., Gulec, M. and Hasde, M. (2011). Analysis of children's rooms in terms of microbiological air quality. *J. Exp. Integr. Med.;* 1: 51-58.

[33] Hargreaves, M., Parappukkaran, S., Morawska, L., Hitchins, J., Congrong, H. and Gilbert, D. (2003). A pilot investigation into associations between indoor airborne

fungal and non-biological particle concentrations in residential houses in Brisbane. *Sci Total Environ*; 312: 89-101.

[34] Heikkienen, M.S.A., Hjelmroos-Koski, M.K., Haggblom, M.M., Macher, J.M. (2005). Bioaerosols. In: Ruzer, L.S., Harley, N.H., Eds. *Aerosols Hand-book.* Boca Raton: CRC Press; pp. 291-342.

[35] Hoskins, J.A., (2007). Health effects due to indoor air pollution. *Indoor Built Env.;* 12: 427-433.

[36] Hyvärinen, A., Meklin, T., Vepsäläinen, A. and Nevalainen, A. (2002). Fungi and actinobacteria in moisture-damaged building materials- concentrations and diversity. *International Biodeterioration and Biodegradation;* 49: 27-37.

[37] Jaenicke, R. (2005). Abundance of cellular material and proteins in the atmosphere. *Science*; 308:73.

[38] Johnson, A.B.R. and Aderele, W.I. (1992). The association of household pollutants and socio-economic risk factors with the short-term outcome of acute lower respiratory infections in Nigerian children. *Annals of Tropical Pediatrics*; 12: 421-432.

[39] Johnson, A.W, Aderele, W.I, Osinusi, K. et al (1996). Acute bronchiolitis in Tropical Africa: a hospital based perspective in Ibadan, Nigeria. *PediatrPulmonol*; 24: 236-247.

[40] Lin, H.H., Murray, M. *et al.* (2008). Effects of smoking and solid-fuel use on COPD, lung cancer, and tuberculosis in China: a time-based, multiple risk factor, modelling study. *Lancet;* 372(9648): 1473-1483.

[41] Lawrence, R. and Martin, D. (2001). Moulds, moisture and microbial contamination of first nations housing in British Columbia, Canada. *International Journal of Circumpolar Health;* 60(2): 150-156.

[42] Li, Y., Leung, G.M., Tang, J.W. and Yang, X. *et al.*, (2007). Role of ventilation in airborne transmission of infectious agents in the built environment-a multidisciplinary systematic review. *Indoor Air;* 17: 02-18.

[43] Manandhar, A., Lee, S.C., Ho, K.F., Hung, W.T. and Joshi, G.R. (2010). Identification of Indoor and Outdoor Particulate Concentration and its Chemical Composition in Kathmandu Valley.

[44] National Health and Medical Research Council (NHMRC) (1996). Definition of indoor air. Retrieved November 2010 from www.nhmrc.gov.au

[45] Ojima, M. *et al.* (2002). Bacterial contamination of Japanese households and related concern about sanitation. *Environmental Health Research;* 12(1): 41-52.

[46] Okafor, E.E., Hassan, A.R. and Hassan, A.D. (2008): Environmental issues and corporate social responsibility: The Nigeria experience. *J. Hum. Ecol.;* 23(2): 101-107.

[47] Oluwole, O., Ana, G.R., Arinola, G.O., Wiskel, T., & Falusi, A.G., Huo, D., Olopade, O.I. and Olopade, C.O. (2013). Effect of stove intervention on household air pollution

and the respiratory health of women and children in rural Nigeria. *Air Qual Atmos Health;* 6: 553-561.

[48] Peppin, J., Demers, A.M., Mberyo-yaah, F., Jaffer, S., Blais, C. and Somse, P. (2001). Acute lower respiratory tract infection among children hospitalized in Bangui, Central African Republic: towards a new case management algorithum. *Trans R Soc Trop Med Hyg*; 95: 410-417.

[49] Perez, M., Ivan, N., Lucia, G., Pruneda, A., Fernando, D., Lilia, E., Batres, E., Francisco, J.,

[50] Perez, V. and Rebeca Isabel, M. S. (2011).Indoor air pollution in

[51] Mexico The Impact of Air Pollution on Health Economy Environment and Agricultural

[52] Sources.

[53] Pillai, S.D. and Ricke, S.C. (2002).Bioaerosols from municipal and animal wastes: background and contemporary issues. *Can J Microbiol*; 48: 681-696.

[54] Pokhrel, A.K., Smith, K.R. *et al.* (2005). Case – control study of indoor cooking smoke exposure and cataract in Nepal and India. *Int J Epidemiol;* 34(3):702-708.

[55] Pudpong, N., Krassi, R. and Nipapun, K. (2011). Indoor concentrations of PM10 and factors influencing its concentrations in day care centres in Bangkok, Thailand. *Asia Journal of Public Health Journal*; 2(1).

[56] Regalado, J., Perez-Padilla, R., Sansores, R., Paramo, R.J.I., Brauer, M., Pare, P. (2006). The effect of biomass burning on respiratory symptom s and lung function in rural Mexican women. *Am RespirCrit Care Med J*;174(8): 901-905.

[57] Romieu, I., Castro-Giner, F., Kunzli, N. and Sunyer, J. (2008). Air pollution, oxidative stress and dietary supplementation: a review. *EurRespir J.;* 31(1):179-197.

[58] Rosenstreich, D.L. et al. (1997). The role of cockroach allergy and exposure to cockroach allergen in causing morbidity among inner-city children with asthma. *New England Journal of Medicine;* 336(19): 1356-1363.

[59] Saha, A., Rao, N.M., *et al.* (2005). Pulmonary function and fuel use: a population survey. *Respir Res.;* 6:127.

[60] Shrestha, I.L. and Shrestha, S.L. (2005). Indoor air pollution from biomass fuels and respiratory health of the exposed population in Nepalese households. *Int J Occup Environ Health;* 11(2):150-160.

[61] Siddiqui, A.R., Lee, K., Gold, E.B., Bhutta, Z.A. (2005). Eyes and respiratory symptoms among women exposed to wood smoke emitted from indoor cooking: a study from southern Pakistan. Energy for Sustainable Development; 9(3):58-66.

[62] Smith, S. J., Alexander, A., Easterlow, D. (1997). Rehousing as a health intervention: miracle or mirage? *Health and Place*; 11: 271-286.

[63] Smith, K.R., Mehta, S., *et al.* (2004). Indoor smoke from household use of solid fuels. In: Ezzati, M., Lopez, AD., Rodgers, A., Murray, C.J.L. (eds) Comparative quantification of health risks: the global burden of disease due to selected risk factors. *World Health Organization*; 2:1435-1493.

[64] Smith-Severtsen, T., Diaz, E., Pope, D. (2009). Effect of reducing indoor air pollution on women's respiratory symptoms and lung function: the RESPIRE Randomized Trial, Guatemala. *Am J Epidemiol*;170(2): 211-220.

[65] Speirs, J.P., Anderson, A. and Anderson, J.G. (1995). A study of microbial content of domestic kitchen. *International Journal of Environmental Health*; 5(13): 109-122.

[66] Srikanth, P., Sudharsanam, S., Steinberg, R. (2008). Bioaerosols in indoor environment: Composition, health effects and analysis. *Indian J Med Microbiol*; 26: 302-312.

[67] Stanley, A.M., Mbamali, I., Zubairu, I.K., Bustani, S.A., Andrew, S.S. & Joshua, I.A. (2010). Electric Power Generator Noise Level Characterization and Impact on Selected Commercial Areas of Zaria and Kaduna Nigeria, International Postgraduate Conference on Infrastructure and Environment, The Hong Kong Polytechnic University, Hong Kong.

[68] Stetzenbach, L.D., Buttner, M.P. and Cruz, P. (2004). Detection and enumeration of airborne contaminants. *Curr Biotechnol*; 15: 170-174.

[69] Strachan, D.P., Cook, D.G. and Derek, G. (1998). Parental smoking and allergic sensitization in children: longitudinal and case-control studies. *Thorax;* 53(3): 204-212.

[70] Toivola, M., Alm, S., Reponen, T., Kolari, S. and Nevalainen, A. (2002). Personal exposures and micro-environmental concentrations of Particles and Bioaerosols. *Journal of Environment Monitoring*; 4:166.

[71] United States Environmental Protection Agency (USEPA) (1997). Revisions to the national ambient air quality standards for particles matter. *Federal Register*; 62: 38651-38701.

[72] Verreault, D., Moineau, S. and Duchaine, C. (2008). Methods for sampling of airborne viruses. Microbiology and Molecular Biology Reviews; 72(3): 413-444.

[73] WHO (2000). The World Health Report 2000 – Health systems: Improving performance. World Health Organization, Geneva.

[74] World Health Report (2002). Reducing risks, promoting healthy life. Geneva, World

[75] Health Organization (WHO). Retrieved March 4, 2010, from http://www.who.int/whr/2003/chapter1/en/index4.html

[76] World Health Organization (2007). Public Health and the Environment. Geneva: WHO.

[77] Yli-Pirilä, T., Kusnetsov, J., Haatainen, S., Hänninen, M., Jalava, P., Rei-man, M. et al. (2004). Amoebae and other protozoa in material samples from moisture-damaged buildings. Environ *Res*; 96(3): 250-256.

[78] Yli-Pirilä, T., Huttunen, K., Nevalainen , A., Seuri, M., Hirvonen, M. (2007). Effects of co-culture of amoebae with indoor microbes on their cytotoxic and pro-inflammatory potential. *Environ Toxicol*; 22(4): 357-367.

[79] Zock, J.P., Heinrich, J., Jarvis, D., Verlato, G., Norback, D., Plana, E. et al. (2006). Distribution and determinants of house dust mite allergens in Europe: the European Community Respiratory Health Survey II. J Allergy ClinImmunol; 118(3): 682-690.

3

On the Analytical Formulations for Pollutant Dispersion Simulation in the Atmospheric Boundary Layer

Davidson Martins Moreira, Tiziano Tirabassi and Taciana Albuquerque

Additional information is available at the end of the chapter

1. Introduction

The management and preservation of air quality require information of the environmental status. Such information involves both cognitive and interpretative features. Networks and measurements for monitoring, together with emission inventory of sources, are of great significance for the building of the cognitive representation, but not the interpretative one. Air quality control demands interpretative tools that are able to extrapolate in space and time the values measured by analytical instrumentation at field sites, while environmental improvement can only be obtained by means of emissions reduction from a systematic planning by using tools (mathematical models of dispersion) adequate to link the causes (sources) with the effects (pollutants).

The pollutants' transport and diffusion processes are complex and without mathematical models it is impossible to account for them. Such models therefore are an indispensable technical instrument of air quality management.

The theoretical approach to the atmospheric dispersion assumes different points of view. The diffusion in the K approach is proportional to the local gradient concentration from diffused material (at the fixed point in space). As a result, it is fundamentally Eulerian, once it assumes the motion of fluid within a fixed system of reference.

Lagrangian models differ from Eulerian ones in adopting a system of reference that follows atmospheric motions. This category contains all models that consider the pollutant cloud as discrete "elements" (puffs or computer particles). In Lagrangian approach, the dispersion is simulated through the motion of particles whose path allows the concentration field computation from the liberated material.

In this chapter, we consider the Eulerian approach.

2. K models

The Eulerian models takes into account the resolution of the mass conservation equation of pollutant chemical species [1]:

$$\frac{\partial c}{\partial t} + u\nabla c = D\nabla^2 c + S. \tag{1}$$

In (1), $\boldsymbol{u}$ is the wind speed vector, $D\nabla^2 c$ is the molecular diffusion, D is the molecular diffusion coefficient, S is the term source, is an operator, and ∇^2 is the Laplacian.

To obtain the solution of equation (1), it is inevitable to know the wind field, something that is not possible because it is changing in space and time. As a consequence, wind is divided into two parts:

$\bar{u}$, ensemble average

u', turbulent fluctuation

Therefore, the wind speed can be expressed as the sum of the two contributions:

$$u = \bar{u} + u' \tag{2}$$

with $\bar{u}'=0$.

Similar hypothesis can be assumed for the concentration:

$$c = \bar{c} + c' \tag{3}$$

with $\overline{c'} = 0$.

The new u and c are introduced into (1). Then, the following is obtained:

$$\frac{\partial \bar{c}}{\partial t} = -\bar{u}\cdot\nabla\bar{c} - \nabla\cdot\overline{c'u'} + D\nabla^2\bar{c} + \bar{S}. \tag{4}$$

This equation possesses new unknown variables. This feature leads to a number of unknowns greater than the number of equations. By solving this problem it is possible to parameterize some terms with known quantities.

The widely used approach is the parameterization of second order moments. Such an approach is called as K-theory or flux-gradient theory:

$$\overline{c'u'} = -K\nabla\overline{c}, \tag{5}$$

where K is the well-known eddy diffusivity.

A great variety of K-formulations exists [2]. The major part them take into account the similarity theory, and give different results for the same atmospheric stability.

Applying the following approximations:

a. Tensor is diagonal

b. c represents the concentration of a non-reactive pollutant (thusS = S)

c. The molecular diffusion is insignificant

Then, equation (4) can be written as:

$$\frac{\partial\overline{c}}{\partial t} = -\overline{u}\cdot\nabla\overline{c} + \nabla\cdot K\nabla\overline{c} + S. \tag{6}$$

Equation (6) can be solved analytically or numerically, with data for u, K, and S and the initial and boundary conditions for$c$. This equation can be resolved in two ways:

i. With analytic methods, obtaining exact solutions

ii. With numerical methods, obtaining approximate solutions

In this chapter we take into consideration analytical solutions.

3. Analytical solutions

Analytical solutions of equations are very important to understand and describe physical phenomena. Moreover, they are fast, simple, and, generally, do not require complex meteorological inputs.

There are analytical solutions of two-dimensional advection-diffusion equation [3]:

$$u\frac{\partial\overline{c}(x,z)}{\partial x} = \frac{\partial}{\partial z}\left(K_z\frac{\partial\overline{c}(x,z)}{\partial z}\right) + S, \tag{7}$$

where u is longitudinal mean speed, $\bar{c}$is the mean concentration, S is the source term, and K_z is the vertical eddy diffusivity. Furthermore, the longitudinal diffusion can be neglected because it is considered less in respect to the advection. Although, very recently a steady-state analytical solution for dispersion of contaminants in low winds by taking into account the longitudinal diffusion in the advection-diffusion equation was formulated [4].

The best-known solution is the so-called Gaussian solution, where both wind and turbulent diffusion coefficients are constant with height. So it is not a realistic solution of the transport and diffusion equation in the atmosphere. In the so-called Gaussian models, the solution is forced to represent real situations by means of empirical parameters, referred to as "sigmas." The different versions of Gaussian models substantially differ in the techniques utilized to calculate the "sigmas" as a function of atmospheric stability and the downwind distance from the emission source. Gaussian models are fast, simple, and do not require complex meteorological inputs. For these reasons they are still widely used by all the environmental agencies over the world for regulatory applications. At this point, it is important to mention that there are models based on non-Gaussian analytical solutions.

[5] report a bi-dimensional solution considering cases where the wind speed and vertical eddy diffusivities follow power laws as a function of height:

$$u=u_1\left(z/z_1\right)^{\alpha} \tag{8}$$

$$K_z=K_1\left(z/z_1\right)^{\beta}, \tag{9}$$

where z_1 is the height when u_1 and K_1 are computed.

[6] report a bi-dimensional solution for linear profiles of the eddy diffusivity. [7] showed the bi-dimensional equation with u and K power functions of height with the exponents following the law of Schmidt.

[8] shows a solution considering u constant, but K as follows:

$$K_z = K_0 z^a \left(h-z\right)^b, \tag{10}$$

where $K_z$is a constant and a and b can be expressed as:

$a^3\ 0 ; b = 0$
$a = 0 ; b > 0$ for $0£ \leq z \leq h$
$a = 1 ; b > 0$ for $0 \leq z \leq h$
$a = 1 ; b = 0$ for $0 \leq z \leq h/2$; $a = 0$ and $b = 1$ for $h/2 \leq z£ \leq h$,

where h is the ABL height.

[9] proposed a solution with constant u and K_z as:

$$K_z \equiv z \quad \text{for } 0 \leq z \leq z_s \tag{11}$$

$$K_z = K_z(z_1) \quad \text{for } z_s < z \leq h, \tag{12}$$

where z_s is a predetermined height (generally, the height of the surface layer). This solution allows (as boundary conditions) a net flow of material toward the ground:

$$K_z \frac{\partial \bar{c}}{\partial z} = V_g \bar{c}, \tag{13}$$

where V_g is the deposition velocity.

[10, 11] report bi-dimensional solutions considering elevated sources with wind speed and vertical eddy diffusivity power profiles, but for an unbound atmosphere. That is:

$$K_z \frac{\partial \bar{c}}{\partial z} = 0 \text{ at } z = \infty. \tag{14}$$

[12] shows a solution considering identical conditions, but for a vertically limited boundary layer. That is:

$$K_z \frac{\partial \bar{c}}{\partial z} = 0 \text{ at } z = h\,. \tag{15}$$

The solutions [10, 11, 12] are used in the KAPPAG model [13, 14, 15].

[16] using the Monin-Obukhov similarity theory to diffusion, has derived a solution from continuous sources near the ground with u and K follow power profiles. SPM [17] is a model that uses this solution.

[18] report a solution, which was a particular case of [7, 8]. In the sequence, [19] extended the solution to the situation of a growing ABL. [20] extended the solution to the case of nonzero mean vertical wind profiles.

[21] presented a three-dimensional atmospheric dispersion-deposition model for a ground-level area source.

[22] extended the solution obtained in [12] with boundary conditions considering dry deposition to the ground.

In the work [23] were reported equations for point sources releases considering the first four moments of the vertical concentration distribution and the magnitude and downwind location of the maximum ground concentration.

[24] obtained a three-dimensional solution (with u and K_z constant) for low wind condition where the diffusion coefficients are function of down-wind distance from the source. [25] presented a two-dimensional solution with K_z constant and function of down-wind distance from the source, with a power low wind profile and for a finite boundary layer.

[26] obtained an analytical solution with dry deposition to the ground with any vertical function of wind and eddy coefficients but for a fixed vertical shape of contaminant concentrations.

$$\bar{c}(x,z) = F(x)\left(1 - \frac{z}{h}\right)^2, \tag{16}$$

where $F(x)$ is any function of x.

Using ADMM (Advection-Diffusion Multilayer Method) approach, [27] proposed a general solution for any wind and eddy coefficient profiles, but represented by a stepwise function in z. Recently, using the ADMM approach, [28] found a solution for two-dimensional steady-state solution considering eddy coefficient profile depending of x and z variables.

Finally, [29, 30] applying GILTT technique (Generalized Integral Laplace Transform Technique), found a general two-dimensional steady-state solution for any profiles of wind and eddy coefficient diffusions and a limited ABL.

3.1. GILTT approach solution

The General Integral Transform Technique (GITT) is a well-known hybrid method that solved a wide class of direct and inverse problems mainly in the area of Heat Transfer and Fluid Mechanics. There is a vast literature about this subject, including papers and books that become impossible to mention all of them. But, for instance, we mention the works of [31, 32, 33, 34], etc. In this chapter, we restrict our attention to the linear problem, because for nonlinear problem the linear result is iteratively used after the linearization of nonlinear transformed equation. Indeed, for the linear problems, the GITT transformed problem is analytically solved by the Laplace Transform technique. Consequently, the GITT solution becomes an analytical solution in the sense that no approximations are made along its derivation. It is important to notice that the GITT solution for nonlinear problems is semi-analytical because, in each iteration, the solution is analytical. This methodology is known as GILTT (Generalized Integral Laplace Transform Technique).

Bearing in mind that the novelty of the GILTT approach relies on the analytical solution of GITT linear transformed problem, in what follows, we restrict our analysis to the ensuing standard GITT time-dependent transformed problem:

$$E\frac{dY(t)}{dt} + FY(t) = 0 \qquad t > 0, \tag{17}$$

subject to the initial condition,

$$Y(0) = Y_0. \tag{18}$$

Here, Y(t) is the unknown vector with N components, E and F are constant matrices of order (NxN), and Y_0 is a known vector with N components. We begin our analysis recasting equation (17) like,

$$\frac{dY(t)}{dt} + GY(t) = 0 \qquad t > 0, \tag{19}$$

where $G = E^{-1}F$ and E^{-1} denotes the inverse of the matrix E. Now, applying the Laplace transform technique to equation (19), we obtain,

$$s\overline{Y(s)} + G\overline{Y(s)} = Y_0, \tag{20}$$

where $\overline{Y(s)}$ denotes the Laplace transform of the vector $Y(t)$. Assuming that the matrix G is nondegenerate, we may write,

$$G = XDX^{-1}. \tag{21}$$

Here, D is the eigenvalue of the diagonal matrix, X is the matrix of the respective eigenfunctions, and X^{-1} its inverse.

Indeed, replacing equation (21) in equation (20) we get,

$$s\overline{Y(s)} + XDX^{-1}\overline{Y(s)} = Y_0, \tag{22}$$

or

$$\left(sI + XDX^{-1}\right)\overline{Y(s)} = Y_0, \tag{23}$$

where the matrix I is the (NxN) identity matrix. Recalling that $X X^{-1} = X^{-1}X = I$, equation (23) reads like,

$$\left(sXX^{-1} + XDX^{-1}\right)\overline{Y(s)} = Y_0, \tag{24}$$

or

$$X\left(sI+D\right)X^{-1}\overline{Y\left(s\right)}=Y_0, \tag{25}$$

which has the well-known solution,

$$\overline{Y\left(s\right)}=X\left(sI+D\right)^{-1}X^{-1}Y_0. \tag{26}$$

Performing the Laplace transform inversion of equation (26), we have,

$$Y\left(t\right)=X\,L^{-1}\left\{\left(sI+D\right)^{-1}\right\}X^{-1}Y_0. \tag{27}$$

Here, L^{-1} denotes the inverse Laplace transform operator. We must notice that the matrix $(sI + D)$ has the form,

$$\left(sI+D\right)=\begin{pmatrix} s+d_1 & 0 & \cdots & 0 \\ 0 & s+d_2 & \cdots & 0 \\ \vdots & \vdots & \ddots & \ddots \\ 0 & 0 & \cdots & s+d_N \end{pmatrix}, \tag{28}$$

where d_i are the eigenvalues of the matrix G. From the diagonal structure of the matrix G, we can straightly write down its inverse like:

$$\left(sI+D\right)^{-1}=\begin{pmatrix} \dfrac{1}{s+d_1} & 0 & \cdots & 0 \\ 0 & \dfrac{1}{s+d_2} & \cdots & 0 \\ \vdots & \vdots & \ddots & \ddots \\ 0 & 0 & \cdots & \dfrac{1}{s+d_N} \end{pmatrix}. \tag{29}$$

Performing the Laplace transform inversion of equation (29) by using the standard results of the Laplace transform theory we obtain:

$$L^{-1}\left\{\left(sI+D\right)^{-1}\right\}=E(t)=\begin{pmatrix} e^{-d_1 t} & 0 & \cdots & 0 \\ 0 & e^{-d_2 t} & \cdots & 0 \\ \vdots & \vdots & \ddots & \ddots \\ 0 & 0 & \cdots & e^{-d_N t} \end{pmatrix}. \tag{30}$$

Finally, substituting the above ansatz in equation (27), we come out with the following solution for problem (17),

$$Y(t) = X \begin{pmatrix} e^{-d_1 t} & 0 & \cdots & 0 \\ 0 & e^{-d_2 t} & \cdots & 0 \\ \vdots & \vdots & \ddots & \ddots \\ 0 & 0 & \cdots & e^{-d_N t} \end{pmatrix} X^{-1} Y_0. \tag{31}$$

To this point, it is relevant to underline that we are aware that problem (17) has a well-known solution. However, we must point out that the discussed solution is a robust algorithm, under computational point of view, to solve the problem with large N (N of order of 1, 500) and a small computational effort. Furthermore, this methodology can also be readily applied for the solution of equation (17) with boundary condition. This kind of a problem appears in the solution of discrete ordinates equation in a slab by the LTS_N approach. For more information, see [35].

Next, we extend this methodology to the solution of more general linear ordinary differential equation having the entries of the matrices E and F varying with time. To reach this goal, let us consider next the problem:

$$E(t)\frac{dY(t)}{dt} + F(t)Y(t) = 0 \qquad t > 0, \tag{32}$$

subject to the initial condition,

$$Y(0) = Y_0. \tag{33}$$

To solve this problem by the Laplace transform technique, we perform a stepwise approximation of the matrices E(t) and F(t). So far, bearing in mind the interest of finding a solution for the time ranging from zero to a prescribed time T, we split the interval (0, T) into sub-interval (t_{n-1}, t_n) for n=1:M, where M denotes the number of sub-intervals. For each sub-interval, we take the averaged values for the entries of matrix E(t) and F(t). It turns out that problem (28) simplifies to the recursive set of problems,

$$E^m \frac{dY^m(t)}{dt} + F^m Y^m(t) = 0, \tag{34}$$

for t in the sub-interval (t_{m-1}, t_m) and m ranging from 1 to M. Here, E_m and F_m are now constant matrices. Henceforth, the previously discussed solution can be applied in a straightforward manner. Indeed, the solution for problem (34) generically reads like,

$$Y^{(m)} = X^{(m)} \begin{pmatrix} e^{-d_1^{(m)}t} & 0 & \cdots & 0 \\ 0 & e^{-d_2^{(m)}t} & \cdots & 0 \\ \vdots & \vdots & \ddots & \ddots \\ 0 & 0 & \cdots & e^{-d_N^{(m)}t} \end{pmatrix} X^{(m)-1} Y^{(m-1)}, \tag{35}$$

for t in the interval (t_{m-1}, t_m) and the initial condition $Y^{(m-1)}$ given by the previously calculated solution at t_{m-1} for t in the interval (t_{m-2}, t_{m-1}).

Focusing our attention on the task of searching analytical solution for the GITT linear transformed equation, in the sequel, we report an analytical alternative approach to solve equation (19), skipping the stepwise approximation of matrices E(t) and F(t) entries. To hit this objective, we solve problem (33) by the decomposition method proposed by [36]. In order to apply the decomposition method, let us recast equation (34) as,

$$\bar{E}\frac{dY(t)}{dt} + \bar{F}Y(t) = A(t)\frac{dY(t)}{dt} + B(t)Y(t), \tag{36}$$

where the matrices A(t) and B(t) are, respectively, expressed as,

$$A(t) = E(t) + \bar{E} \tag{37}$$

and

$$B(t) = F(t) + \bar{F}. \tag{38}$$

Here, $\bar{E}$ and $\bar{F}$ are constant matrices properly chosen. Now, expanding the solution of equation (36) in the truncated series,

$$Y(t) = \sum_{k=0}^{L} u_k(t) \tag{39}$$

and replacing this ansatz in equation (36), we get,

$$\bar{E}\frac{d}{dt}\sum_{k=0}^{L} u_k(t) + \bar{F}\sum_{k=0}^{L} u_k(t) = A(t)\frac{d}{dt}\sum_{k=0}^{L} u_k(t) + B(t)\sum_{k=0}^{L} u_k(t). \tag{40}$$

From equation (40), we are in a position to construct the following recursive set of linear ordinary differential matrix equations with constant matrices,

$$\bar{E}\frac{du_0(t)}{dt} + \bar{F}u_0(t) = 0, \tag{41}$$

$$\bar{E}\frac{du_1(t)}{dt} + \bar{F}u_1(t) = A(t)\frac{du_0(t)}{dt} + B(t)u_0(t), \tag{42}$$

$$\bar{E}\frac{du_2(t)}{dt} + \bar{F}u_2(t) = A(t)\frac{du_1(t)}{dt} + B(t)u_1(t) \tag{43}$$

and so forth. Generically, we may write,

$$\bar{E}\frac{du_n(t)}{dt} + \bar{F}u_n(t) = A(t)\frac{du_{n-1}(t)}{dt} + B(t)u_{n-1}(t), \tag{44}$$

for n = 1:L. Given a closer look at the above equations, we readily realize that the RHS of the recursive set of equations from (42) to (44) are known terms. Therefore, to obtain the solution of these equations, we must solve the following problem with constant matrices and source, namely,

$$\bar{E}\frac{dY(t)}{dt} + \bar{F}Y(t) = S(t) \qquad t > 0, \tag{45}$$

which has the well-known solution,

$$Y(t) = Y_h(t)Y_0 + Y_h(t) * S(t), \tag{46}$$

whereas the homogeneous solution $Y_h(t)$ is given by equation (35). Here, star denotes convolution. Therefore, the solution of the generic equation (44) has the form,

$$Y(t) = X\begin{pmatrix} e^{-d_1 t} & 0 & \cdots & 0 \\ 0 & e^{-d_2 t} & \cdots & 0 \\ \vdots & \vdots & \ddots & \ddots \\ 0 & 0 & \cdots & e^{-d_N t} \end{pmatrix} X^{-1}Y_0 + \\ + \int_0^t \left(X(t')\begin{pmatrix} e^{-d_1 t'} & 0 & \cdots & 0 \\ 0 & e^{-d_2 t'} & \cdots & 0 \\ \vdots & \vdots & \ddots & \ddots \\ 0 & 0 & \cdots & e^{-d_N t'} \end{pmatrix} X(t')^{-1} S(t-t') \right) dt' \tag{47}$$

and consequently the solution of equation (32) is well determined by equations (39) and (41) to (44).

Finally, regarding the task of solving second-order ordinary differential transformed equation, we must recall that this equation can be transformed into a set of first-order matrix differential equation by changing the variable. Indeed, performing the substitution of,

$$Y(t) = Z_1(t) \tag{48}$$

and

$$\frac{dY(t)}{dt} = Z_2(t), \tag{49}$$

in the following transformed problem:

$$\frac{d^2Y(t)}{dt} + A(t)\frac{dY(t)}{dt} + B(t)\,Y(t) = S(t), \tag{50}$$

we come out with the ensuing result:

$$\frac{d}{dt}\begin{pmatrix} Z_1(t) \\ Z_2(t) \end{pmatrix} + \begin{pmatrix} 0 & -1 \\ B(t) & A(t) \end{pmatrix}\begin{pmatrix} Z_1(t) \\ Z_2(t) \end{pmatrix} = \begin{pmatrix} 0 \\ S(t) \end{pmatrix}, \tag{51}$$

which is a set of first-order ordinary matrix differential equation that can be promptly solved by the previous discussed methods for nondegenerated matrix. Finally, to handle transformed problems with degenerated matrix, we proceed by performing the inversion of the symbolic matrix by the Schur decomposition approach. Furthermore, we make the Laplace transform inversion applying the Heaviside expansion formulation for eigenvalues with multiplicity larger than one. Hoping that we have completed the mathematical analysis regarding the task of solving analytically the GITT transformed problems, in the next section, we illustrate the GILTT aptness to simulate pollutant dispersion in the atmosphere by solving the two-dimensional steady-state advection-diffusion equation.

3.1.1. An example: Solution of the 2D advection-diffusion equation

To solve the problem (7) by the GILTT method we recast equation (7) as ($S = 0$):

$$u(z)\frac{\partial \overline{c}}{\partial x} = K_z(z)\frac{\partial^2 \overline{c}}{\partial z^2} + \left(\frac{\partial K_z(z)}{\partial z}\right)\frac{\partial \overline{c}}{\partial z} \tag{52}$$

with the following boundary conditions:

$$K_z \frac{\partial \overline{c}}{\partial x} = 0 \text{ at z = h} \tag{53}$$

$$u\overline{c}(x,z) = Q\delta(z - H_s) \text{ at x = 0} \tag{54}$$

and we construct the following associated Sturm-Liouville problem:

$$\Psi_i''(z) + \lambda_i^2 \Psi_i(z) = 0 \text{ at } 0 < z < h \tag{55}$$

$$\Psi_i'(z) = 0 \text{ at z = 0,h}, \tag{56}$$

which has the well-known solution:

$$\Psi_i(z) = \cos(\lambda_i z), \tag{57}$$

where the eigenvalues λ_i are given by $\lambda_i = \frac{i\pi}{h}$ for $i = 0, 1, 2, 3, \ldots$.Furthermore, The eigenfunctions $\Psi_i(z)$ satisfy the ensuing orthonormality condition:

$$\frac{1}{N_m^{1/2} N_n^{1/2}} \int_v \Psi_m(z) \Psi_n(z) dv = \begin{cases} 0, \text{ m} \neq \text{n} \\ 1, \text{ m = n} \end{cases}, \tag{58}$$

where N_m is expressed by:

$$N_m = \int_v \Psi_m^2(z) dv. \tag{59}$$

Having solved the Sturm-Liouville problem, we are in a position to construct the GILTT transform formula, which has the form:

$$\overline{c}(x,z) = \sum_{i=0}^{\infty} \frac{\overline{c_i(x)\Psi_i(z)}}{N_i^{1/2}}. \tag{60}$$

Now, replacing the above ansatz in equation (52) we have,

$$u\sum_{i=0}^{\infty}\frac{\overline{\overline{c_i'(x)\Psi_i(z)}}}{N_i^{1/2}}=K_z\sum_{i=0}^{\infty}\frac{\overline{\overline{c_i(x)\Psi_i''(z)}}}{N_i^{1/2}}+\left(\frac{\partial K_z}{\partial z}\right)\sum_{i=0}^{\infty}\frac{\overline{\overline{c_i(x)\Psi_i'(z)}}}{N_i^{1/2}}. \tag{61}$$

Here, we adopt the prime notation for the first derivative and double the prime notation for the second derivative. Multiplying equation (61) by $\frac{\Psi_j(z)}{N_j^{1/2}}$ and integrating from z equal zero to h, we read:

$$-\sum_{i=0}^{\infty}\frac{\overline{\overline{c_i'(x)\lambda}}}{N_i^{1/2}N_j^{1/2}}\int_0^h u\Psi_i\Psi_j\,dz-\sum_{i=0}^{\infty}\frac{\overline{\overline{c_i(x)}}\,\lambda_i^2}{N_i^{1/2}N_j^{1/2}}\int_0^h K_z\Psi_i\Psi_j\,dz+$$
$$+\sum_{i=0}^{\infty}\frac{\overline{\overline{c_i(x)}}}{N_i^{1/2}N_j^{1/2}}\int_0^h\left(\frac{\partial K_z}{\partial z}\right)\Psi_i'\,\Psi_j\,dz=0 \tag{62}$$

for $j=0,\ 1,\ 2,\ \ldots,\ N$. Rewriting equation (62) in matrix fashion, we obtain,

$$Y'(x)+FY(x)=0, \tag{63}$$

where Y(x) is the column vector whose components are $c_j(x)$ and the matrix F is defined as $F=B^{-1}E$. Here, the entries of matrices B and E are, respectively, given by:

$$b_{i,j}=\int_0^h u\Psi_i\Psi_j dz \tag{64}$$

and

$$e_{i,j}=\int_0^h\frac{\partial K_z}{\partial z}\Psi_i'\Psi_j dz-\lambda_i^2\int_0^h K_z\Psi_i\Psi_j dz. \tag{65}$$

Transforming the boundary condition (54) by a similar procedure, we mean, multiplying the boundary condition (54) by $\frac{\Psi_j(z)}{N_j^{1/2}}$ and integrating from z = 0 to h, we obtain:

$$\int_0^h\sum_{i=0}^{\infty}\frac{\overline{\overline{c_i(0)}}\;u\Psi_i\Psi_j}{N_i^{1/2}N_j^{1/2}}dz=\int_0^h\frac{Q\;\delta\;(z-H_s)\Psi_j}{N_j^{1/2}}dz. \tag{66}$$

Using the orthonormality property of the eigenfunctions $\Psi_j(z)$, and the integration property of the generalized delta function, we get the following transformed boundary condition:

$$c_0(0) = \frac{Q\Psi_0(H_s)\sqrt{h}}{u(z)\Psi_0^2(z)dz} \tag{67}$$

for j = 0, and

$$c_i(0) = \frac{Q\Psi_j(H_s)\sqrt{h/2}}{\int_0^h u(z)\Psi_j^2(z)dz} \tag{68}$$

for j ranging from 1 to N. To this point, we are in a position to write down the solution of equation (39), subjected to the boundary condition given by equations (67) and (68), using the results discussed in the previous section. Indeed, the solution writes like,

$$\bar{c}(x,z) = \sum_{i=0}^{\infty} \frac{\overline{c_i(x)}\Psi_i(z)}{N_i^{1/2}}. \tag{69}$$

Here, $c_i(x)$ are the N components of the vector Y(x) given by,

$$Y(x) = Z\begin{pmatrix} e^{-f_1 x} & 0 & \cdots & 0 \\ 0 & e^{-f_2 x} & \cdots & 0 \\ \vdots & \vdots & \ddots & \vdots \\ 0 & 0 & \cdots & e^{-f_N x} \end{pmatrix} G(x)Z^{-1}Y(0), \tag{70}$$

where f_k, for k ranging from 1 to N, are the eigenvalues of the matrix of eigenfunctions and Z^{-1} its inverse. For more details and recent developments, see [29, 30, 37, 38, 39, 40, 41].

4. Remarks

To deal with more realistic situations we need to change to numerical methods. However, it is helpful to check firstly possible analytical solutions in order to obtain a known framework and test solutions. The analytical solutions are efficient for many applications, for example: supply analysis of contaminant scenarios, leading sensitivity analyses to investigate the effects of parameters included in contaminant transport, extrapolation over large times and distances where numerical solutions may be impractical, as screening models for transport processes that cannot be solved exactly, and for validating numerical solutions.

Today, air pollution problems are not treated in the manner described in the present chapter. There are various air pollution situations that require the use of complex mesoscale models to properly describe the dispersion processes and properly represent the relevant chemistry and emission processes. Complex models, such as the CMAQ model (the Community Multiscale Air Quality model), have been designed to simulate air quality by including state of the art techniques for modeling multiple air quality issues. However, in complex models, increasingly more processes, such as sea breeze circulations, urban heat islands, and waves, are represented. Therefore, these models are often perceived as "closed" that cannot without difficulty or effort report the influence of individual processes on air quality. Finally, for many scientific applications, analytical solutions have utility in understanding air dispersion phenomena and some air chemistry phenomena, showing their usefulness in environmental management.

Author details

Davidson Martins Moreira[1*], Tiziano Tirabassi[2] and Taciana Albuquerque[3]

*Address all correspondence to: davidson.moreira@fieb.org.br

1 Center for Integrated Manufacturing and Technology (CIMATEC), Salvador, Brazil

2 Institute ISAC of CNR, Bologna, Italy

3 Universidade Federal de Minas Gerais, Belo Horizonte, Brazil

References

[1] Arya, S. 1999. Air pollution meteorology and dispersion. Oxford University Press, New York, 310 pp.

[2] Seinfeld, J.H., and Pandis, S.N. 1998. Atmospheric chemistry and physics. John Wiley & Sons, New York, 1326 pp.

[3] Tirabassi, T. 2003. Operational advanced air pollution modeling. PAGEOPH 160(1–2), 5–16.

[4] Moreira, D.M., Tirabassi, T., and Carvalho, J.C. 2005a. Plume dispersion simulation in low wind conditions in stable and convective boundary layers. Atmos. Environ. 39(20), 3643–3650.

[5] Roberts, O.F.T. 1923. The theoretical scattering of smoke in a turbulent atmosphere. Proc. Roy. Soc. 104, 640–648.

[6] Rounds, W. 1955. Solutions of the two-dimensional diffusion equation. Trans. Am. Geophys. Union 36, 395–405.

[7] Smith, F.B. 1957a. The diffusion of smoke from a continuous elevated point source into a turbulent atmosphere. J. Fluid Mech. 2, 49–76.

[8] Smith, F.B. 1957b. Convection-diffusion processes below a stable layer. Meteorological Research Committee, N. 1048 and 10739, London.

[9] Scriven, R.A., and Fisher, B.A. 1975. The long range transport of airborne material and its removal by deposition and washout-II. The effect of turbulent diffusion. Atmos. Environ. 9, 59–69.

[10] Yeh, G.T., and Huang, C.H. 1975. Three-dimensional air pollutant modelling in the lower atmosphere. Bound. Layer Meteor. 9, 381–390.

[11] Berlyand, M.Y. 1975. Contemporary problems of atmospheric diffusion and pollution of the atmosphere. Translated version by NERC, USEPA, Raleigh, NC.

[12] Demuth, C. 1978. A contribution to the analytical steady solution of the diffusion equation for line sources. Atmos. Environ. 12, 1255–1258.

[13] Tagliazucca, M., Nanni, T., and Tirabassi, T. 1985. An analytical dispersion model for sources in the surface layer, Nuovo Cimento 8C, 771–781.

[14] Tirabassi, T., Tagliazucca, M., and Zannetti, P. 1986. KAPPA-G, a non-Gaussian plume dispersion model: description and evaluation against tracer measurements. JAPCA 36, 592–596.

[15] Tirabassi, T. 1989. Analytical air pollution advection and diffusion models. Water, Air Soil Poll. 47, 19–24.

[16] Van Ulden, A.P. 1978. Simple estimates for vertical diffusion from sources near the ground. Atmos. Environ. 12, 2125–2129.

[17] Tirabassi, T., and Rizza, U. 1995. A practical model for the dispersion of skewed puffs. J. Appl. Meteor. 34, 989–993.

[18] Nieuwsadt, F.T.M. 1980. An analytical solution of the time-dependent, one-dimensional diffusion equation in the atmospheric boundary layer. Atmos. Environ. 14, 1361–1364.

[19] Nieuwstadt, F.T.M., and de Haan, B.J. 1981. An analytical solution of one-dimensional diffusion equation in a nonstationary boundary layer with an application to inversion rise fumigation. Atmos. Environ. 15, 845–851.

[20] Catalano G.D. 1982. An analytical solution to the turbulent diffusion equation with mean vertical wind. In: Proceedings of the 16th Southeastern Seminar on Thermal Sciences, Miami, 19–21 April 1982, 143–151.

[21] Chrysikopoulos, C.V., Hildemann, L.M., and Roberts, P.V. 1992. A three-dimensional atmospheric dispersion-deposition model for emissions from a ground-level area source. Atmos. Environ. 26, 747–757.

[22] Lin, J.S., and Hildemann, L.M. 1997. A generalised mathematical scheme to analytically solve the atmospheric diffusion equation with dry deposition. Atmos. Environ. 31, 59–71.

[23] Brown, M.J., Arya, S.P., and Snyder, W. 1997. Plume descriptors from a non-gaussian concentration model. Atmos. Environ. 31, 183–189.

[24] Sharan, M., and Yadav, A.K. 1998. Simulation of diffusion experiments under light wind, stable conditions by a variable K-theory model. Atmos. Environ. 32, 3481–3492.

[25] Sharan, M., and Modani, M. 2006. A two-dimensional analytical model for the dispersion of air-pollutants in the atmosphere with a capping inversion. Atmos. Environ. 40, 3479–3489.

[26] Essa, K.S.M., Etman, S.M., and Embaby, M. 2007. New analytical solution of the dispersion equation. Atmos. Res. 84, 337–344.

[27] Vilhena, M.T., Rizza, U., Degrazia, G.A., Mangia, C., Moreira, D.M., and Tirabassi, T. 1998. An analytical air pollution model: development and evaluation. Contr. Atmos. Phys. 71, 315–320.

[28] Moreira, D.M., Moraes, A.C., Goulart, A.G., and Albuquerque, T.T. 2014. A contribution to solve the atmospheric diffusion equation with eddy diffusivity depending on source distance. Atmos. Environ. 83, 254–259.

[29] Wortmann, S., Vilhena, M.T., Moreira, D.M., and Buske, D. 2005. A new analytical approach to simulate the pollutant dispersion in the PBL. Atmos. Environ. 39, 2171–2178.

[30] Moreira, D. M., Vilhena, M.T., Tirabassi, T., Buske, D., and Cotta, R.M. 2005b. Near source atmospheric pollutant dispersion using the new GILTT method. Atmos. Environ. 39, 6289–6294.

[31] Cotta, R.M. 1993. Integral transforms in computational heat and fluid flow. CRC Press, Boca Raton.

[32] Cotta, R.M., and Mikhaylov, M. 1997. Heat conduction lumped analysis, integral transforms, symbolic computation. John Wiley & Sons, Chichester.

[33] Cheroto, S., Mikhailov, M.D., Kakaç, S., and Cotta, R.M. 1999. periodic laminar forced convection: solution via symbolic computation and integral transforms. Int. J. Therm. Sci. 38, 613–621.

[34] Alves, L.S., Cotta, R.M., and Pontes, J. 2002. Stability analysis of natural convection in porous cavities through integral transforms. Int. J. Heat Mass Transfer 45, 1185–1195.

[35] Segatto, C.F., and Vilhena, M.T. 1999. The state-of-the-art of the LTS_N method, mathematics and computation, reactor physics and environmental analysis in nuclear applications – International Conference, Madrid, 2, 1618–1631.

[36] Adomian, G. 1988. A review of the decomposition method in applied mathematics. J. Math. Anal. Appl. 1(135), 501–544.

[37] Buske, D., Vilhena, M.T., Moreira, D.M., and Tirabassi, T. 2007a. Simulation of pollutant dispersion for low wind conditions in stable and convective planetary boundary layer. Atmos. Environ. 41, 5496–5501.

[38] Buske, D., Vilhena, M.T., Moreira, D.M., and Tirabassi T. 2007b. An analytical solution of the advection-diffusion equation considering non-local turbulence closure. Environ. Fluid Mech. 7, 43–54.

[39] Moreira, D.M., Vilhena, M.T., Buske, D., and Tirabassi, T. 2009. The state-of-art of the GLTT method to simulate pollutant dispersion in the atmosphere. Atmos. Res. 92, 1–17.

[40] Tirabassi, T., Buske, D., Moreira, D.M., and Vilhena, M.T. 2008. A two-dimensional solution of the advection-diffusion equation with dry deposition to the ground. JAMC 47, 2096–2104.

[41] Tirabassi, T., Tiesi, A., Buske, D., Vilhena, M.T., and Moreira, D.M. 2009. Some characteristics of a plume from a point source based on analytical solution of the two-dimensional advection-diffusion equation. Atmos. Environ. 43, 2223–2229.

4

A Comprehensive Research of Atmospheric Haze by Optical Remote Sensing in Central China Region (CCR)

Wenqing Liu, Zhenyi Chen, Pinhua Xie,
Jianguo Liu and Tianshu Zhang

Additional information is available at the end of the chapter

1. Introduction

With the changing pollution scenario, the task of improving air quality is now faced with increasingly serious challenges. Aerosols and gaseous pollutants have great impact on the radiative balance of the Earth, and thus on climatic changes. Furthermore, the radiative forcing of natural and anthropogenic aerosols exhibits a strong seasonal and geographical variability, which can affect the diffusion of atmospheric pollution locally [1]. The increase of aerosols or gaseous pollutants results in impairment of visibility. Besides, high aerosol concentration is deleterious to human health. The Chinese Academy on Environmental Planning considers that air pollution is responsible for 411,000 premature deaths in China in 2003, primarily from respiratory problems and heart diseases [2]. Therefore, obtaining information on the vertical distribution of aerosols and gaseous pollutants in the pollution haze is crucial to understanding their optical characteristics as well as their impact on air quality and on human health.

During the past 10 years, with its dense population, rapid economic growth, and dramatic rate of urbanization, China has been experiencing elevated levels of air pollution [3-5]. This is particularly so in the case of the Pearl River Delta (PRD), Yangtze River Delta (YRD), and Central China Region (CCR). Among these, the CCR region is where the frequency of occurrence of haze has been increasing rapidly from 2012. In CCR, NOx, SO_2 and VOCs discharged by pollution sources are the significant primary pollutants, which greatly influence the formation of particulate matter and ozone because of photochemical reaction. Besides, the prevalence of mild and humid climate throughout the year in CCR leads to transformation and accumulation of particulate matter and gaseous pollutants. In addition, regional transportation contributes to combined pollution problems of atmosphere, manifested by high-concentration fine particulate matter, low visibility, high-concentration NO_2 column, etc.

Thus far, extensive studies using different methods have been conducted in China on atmospheric pollution [6-9]. For a better understanding of aerosol properties and their effect on environment, two major aerosol experiments were conducted in mainland China since 2004: (i) East Asian Study of Tropospheric Aerosols -An International Regional Experiment (EAST-AIRE) [10] and (ii) East Asian Study of Tropospheric Aerosols and Impact on Regional Climate (EAST-AIRC) [11]. Information on the properties of aerosols and their temporal and spatial distribution was obtained from studies using both short-term intensive field experiments and long-term operations, as well as modeling studies. Besides these filed experiments, lidar (LIght Ranging and Detection) and DOAS (Differential Absorption Spectroscopy) studies were conducted at several sites in China[12-18]. However, knowledge on the properties of aerosols and trace gases, particularly in the vertical dimension, in CCR China is rather scanty. Also, because of lack of real-time data on the quantum of pollutants discharged by major industrial parks, it is hard to resolve the task of pollution prevention. The present research was designed to take advantage of advanced environmental monitoring technical equipment (including Lidar[19,20], multi-axis DOAS[21,22], vehicle-mounted DOAS /infrared SOF[23-25], and meteorological parameter analyzer, etc.) to establish a stereoscopic observing system for key sources of regional pollution in CCR, China. The campaign was carried out from Dec 2013 to April 2014 in three key cities, namely Anyang (AY), Jiaozuo (JZ) and Sanmenxia (SMX). Continuous and automatic stereoscopic monitoring of primary pollution sources, transportation of pollutants, emission characteristics of pollution from major industrial parks, as well as the transport conditions in the surrounding regions of pollution under specific meteorological conditions was carried out to obtain data that can help in air quality improvement and in development of forecast warning model, besides making up for the deficiency in current ground monitoring system.

The scope of research resolves mainly into two parts:

1. Real-time monitoring of spatio-temporal distribution of regional haze pollution:

Representative monitoring points were identified in selected cities and surrounding areas along the direction of pollution transportation to measure spatial and temporal distribution characteristics of SO_2, NO_2 and aerosol particulate matter with lidar, MAX-DOAS, etc.

2. Real-time monitoring of disorganized pollutant discharge in key industrial park sources:

VOC is an important precursor to the formation of photochemical smog. The type and volume of VOC discharge, as also of SO_2 and NO_2, play a significant role in pollution. Aiming at the key source areas of elevated discharge, vehicle-mounted DOAS and vehicle-mounted SOF-FTIR were utilized for mobile scanning measurement of discharged smoke plumes. Using the absorption characteristics of pollution gas molecules in infrared and ultraviolet spectrum, the concentration of exhaust gas was qualitatively and quantitatively determined through SOF-FTIR and DOAS spectrum analytical methods.

2. Instrument and system

2.1. Lidar

The Lidar system, used for the study of spatial and temporal distribution of atmospheric particles, is shown in Fig 1; it adopts 532m Mie scattering and 532 polarization channels. Laser transmitter launches linear polarization lasers at 532nm by shooting vertically into atmosphere through beam expander, inversed mirror and quartz skylight. The laser pulse is scattered through air molecules, atmospheric aerosols or cloud, on its transmission route. The backward-scattered light firstly passes through skylight glass and then as parallel light through the diaphragm hole and convergent lens of the telescope. The converged parallel light is split into two beams by dichroic mirror: one is at 532nm, received by photomultiplier C after passing through optical filter; the other is at 355nm, whose vertical and parallel components are received respectively by A and B photomultipliers, after decomposition by an analyzer prism. After the conversion of optical signal into electrical signal by photomultiplier, the signal will be delivered to data acquisition system, where its ASCII code data file will be recorded and saved by main control computer.

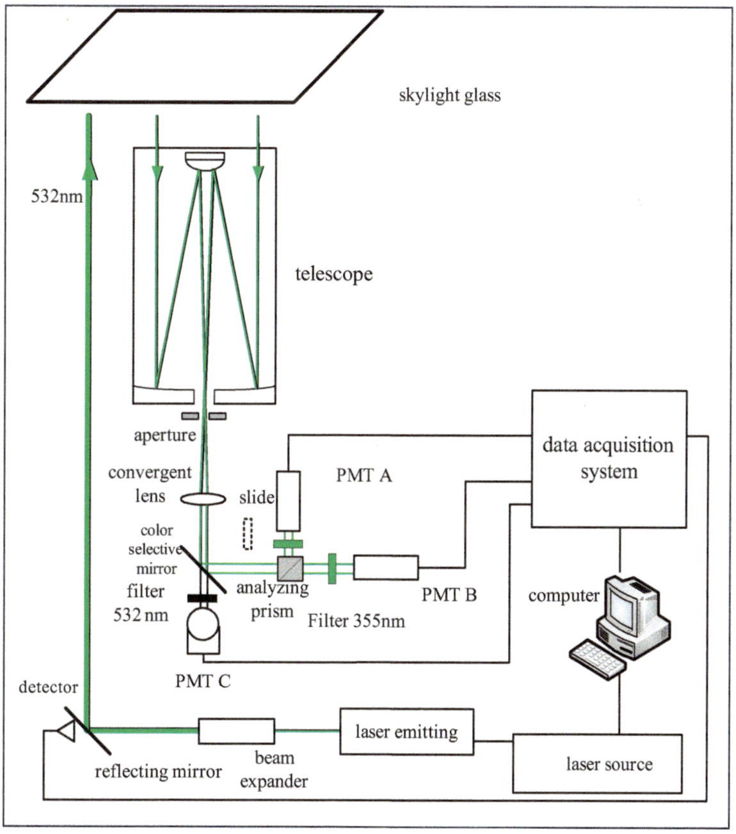

Figure 1. System structure of lidar

2.2. MAX-DOAS

MAX-DOAS is a special type of DOAS technology, which uses characteristic absorption structure of trace gases (such as NO_2, SO_2, O_3) in visible ultraviolet band for qualitative and quantitative analysis of gas. As a real-time and online measuring method, it is eminently suitable for remote sensing measurement of air pollution. Besides, the data on vertical column density of pollution gases acquired from this method can help in verifying satellite results [26]. MAX-DOAS receives scattered sunlight from telescopes at different elevations, as shown in Fig 2. Based on the characteristic absorption spectrums of pollution gas, it inverses the vertical column density of pollution gas and distribution profile of density. Besides, it acquires data on spatial distribution of pollution gas in atmosphere by measuring atmospheric absorption spectrums in zenith direction and several off-axis directions. During the monitoring process, the reflector orients itself to the north and is then driven by stepper motor to turn to 1°, 2°, 3°, 4°... 90° to finish one measurement cycle, thus realizing measurement of different angles. The schematic diagram of MAX-DOAS is presented below.

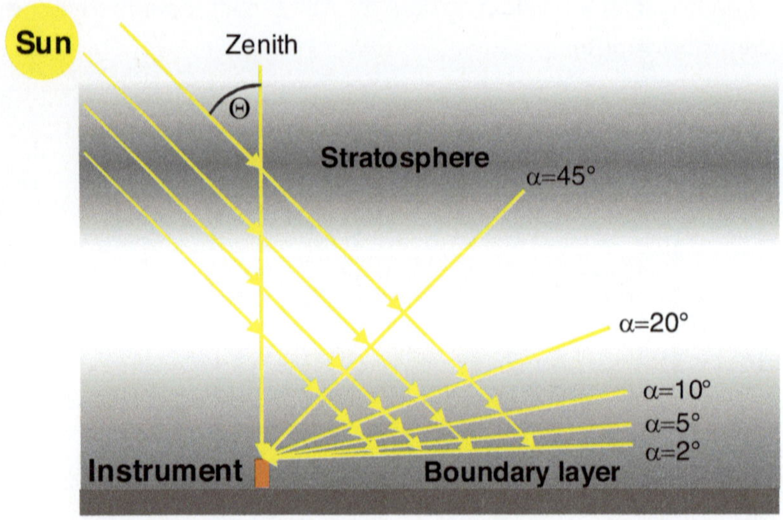

Figure 2. Schematic diagram of MAX-DOAS

2.3. Vehicle-mounted DOAS

With zenith-sky scattered light as the light source, vehicle-mounted DOAS technology is a kind of spectrum detection technology, which uses "fingerprint" absorption characteristics of optical radiation by gas molecule for qualitative and quantitative measurements of gas. The system includes mainly a spectrograph, a GPS system and a telescope [27]. As shown in Fig 3, the whole system is put on mobile platform and the telescope gathers solar spectrum along the measurement route. During operation, the band of spectrograph is fixed at 290nm~420nm (spectral region 130nm) as this band has strong SO_2 and NO_2 absorption structures. After the spectrograph delivers the gathered spectrum to computer for data processing, DOAS method is used to calculate the column density of pollution gas. Finally, density distribution of pollution gas on the measuring route can be acquired by combing the information from GPS on longitude and latitude.

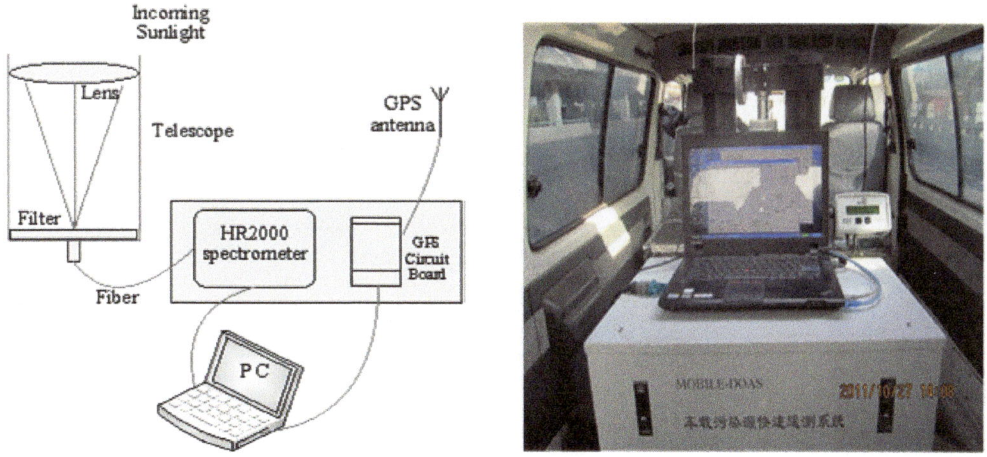

Figure 3. Structure and system of vehicle-mounted DOAS

2.4. Satellite

The spatial resolution of the OMI detector mounted on AURA satellite is 13 km×24 km [28]. NO_2 VCD of troposphere, below the satellite orbit, can be obtained through DOAS inversion of measured spectrum. After differential and meshing treatment, NO_2 VCD of each acquired orbit is integrated and then NO_2 density distribution over the region is obtained. Acquisition of regional density distribution data is significant for investigation of the processes of generation and transportation of pollution. For the present study, OMI Level 2 NO_2 VCD data was adopted. From the weekly average between Nov and Dec, it was found that CCR pollution reached its peak in the third week.

3. Results and discussion

3.1. Stereoscopic monitoring of regular pollution gas

1. Comparison of areal distribution at the three sites

The average of the values obtained during measurement period at the three observation sites (AY, JZ and SMX) were taken as the areal distribution level of NO_2 and SO_2 in CCR (Fig 4). It can be observed that the average level of NO_2 density at AY is $63.4*10^{15}$molec./cm^2, which is higher than the values at JZ and SMX. Density of NO_2 in SMX is the lowest at its average level of $35.3*10^{15}$molec./cm^2. SO_2, density at JZ is the highest at $149*10^{15}$molec./cm^2, which is higher than that at AY and SMX.

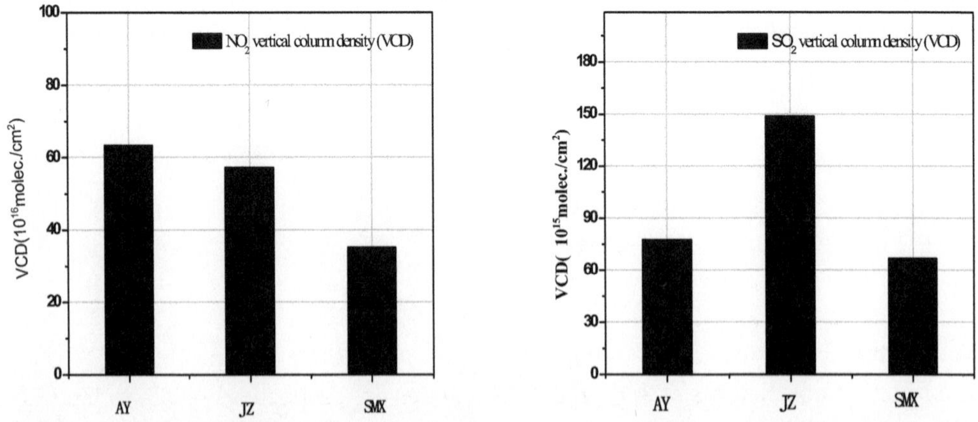

Figure 4. Characteristic contrasts about VCD (vertical column density) of SO_2 and NO_2 at the three sites

2. Results of monthly average variation

The results obtained during the measurement period were used to calculate the averages of different months and to analyze variation trends of pollutant concentration. It can be observed from the calculation results (see Fig 5) that NO_2 VCD (vertical column density) reached its peak in Dec and Jan, decreased from Feb onward, and then remained low between Mar and June. Monthly average of SO_2 VCD at each station is shown in the figure below. It is obvious from VCD of SO_2 is high in winter just as NO_2; it reached its peak in Dec at AY, in Jan at SMX, and in Feb at JZ. It obviously decreased from Mar and thereafter remained low up to June.

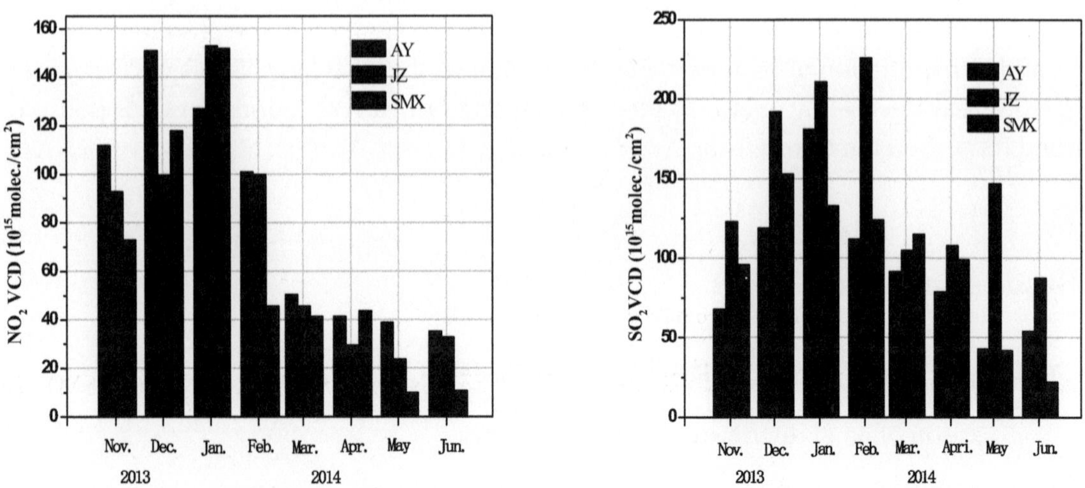

Figure 5. Monthly average variations about VCD (vertical column density) of SO_2 and NO_2 in observing sites

From an overview of the monthly averages discussed above, it can be concluded that the accumulation of pollutants had been accumulated rapidly and heavily because of high discharge and unfavorable weather conditions in winter. With the end of hot season after

February, change of wind direction and increase in rainfall would start favoring diffusion and sedimentation of pollutants, and consequently reduction of pollutant concentration.

3.2. Stereoscopic monitoring of aerosols

Monitoring of aerosols through lidar, from Nov 2013 to April 2014, (Fig 6) revealed that haze pollution was severe mainly from Nov. 2013 to Jan. 2014. Particularly from Dec 12 to Dec 20, and Jan 4 to Jan 17, the pollution lasted for almost half a month at AY and JZ. Overall air quality improved continuously, from Feb to June 2014, at AY, JZ and SMX. However, during wheat harvesting in June, because of straw burning, pollution levels increased at AY, JZ and SMX sites. The highest pollution levels were confined to the middle of June (14th-15th); the peak of aerosol extinction coefficients at AY, JZ and SMX sites are respectively $0.83 km^{-1}$, $0.91 km^{-1}$ and $0.66 km^{-1}$.

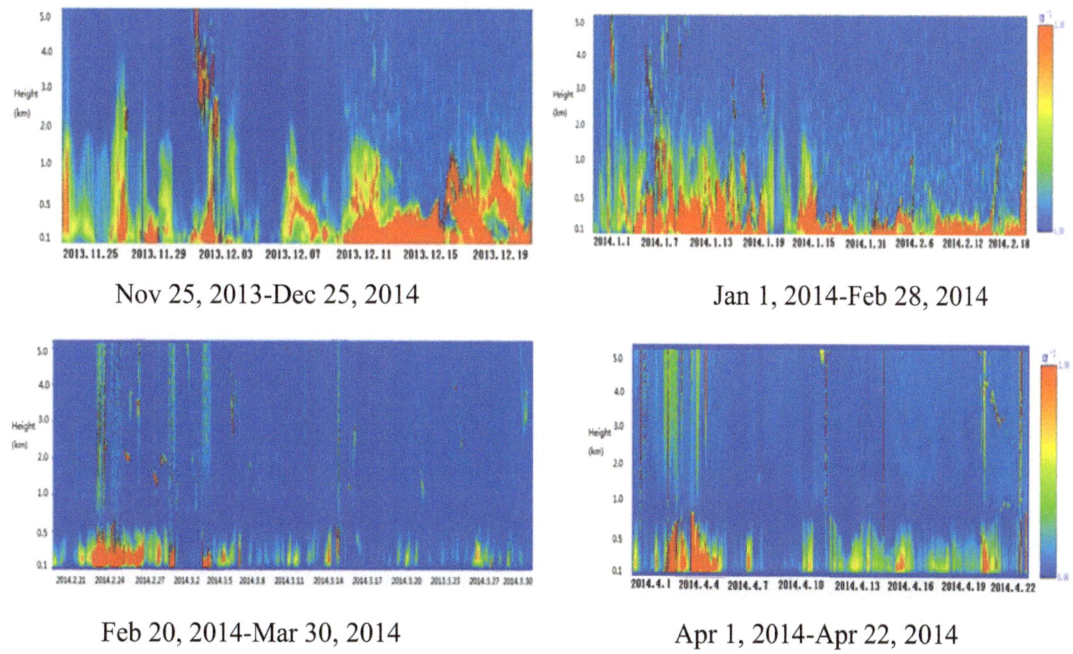

Figure 6. Monthly aerosol distributions from lidar observation

Pollution transportation (Nov 25, 2013-Nov 27, 2013) also occurred at AY, JZ and SMX, but the pollution characteristics and sources were different. Among these three sites, sand dust transportation in northwest direction affected SMX the most. Because of sand dust transportation, large particle concentration in air increased and RH decreased rapidly (at SMX, RH=78% on Dec 25 and RH=32% on Dec 27). Repeated monitoring of local pollutions (Dec 12, 2013-Dec 17, 2013; Jan 8, 2014-Jan 11, 2014; Jan 27, 2014-Jan 31, 2014; Feb 16, 2014-Feb 18, 2014; Apr 10, 2014-Apr 13, 2014) reveal that near-ground pollution at these three sites came mainly from local pollution (industrial emission, motor vehicle exhaust and coal combustion) and fire coal transportation from southern cities in the southeast direction. By the end of Jan 2014, local pollution aggravated because of discharge from firecrackers during spring festival. The pollution at higher levels in the air above 1km came mainly from the northwest.

In conclusion, it emerges that particulate pollution occurred almost simultaneously at the three sites AY, JZ and SMX. Local pollution (local source) is typical in AY, and near-ground pollution in JZ is higher due to the local industrial emission. Local pollution was relatively slighter in SMX, but it can be easily aggravated by northwest sand dust transportation. Sand dust pollution at AY manifests as superposed locally raised dust over northwest sand dust. In CCR, the ground weather conditions of uniform-pressure were conducive to pollutant diffusion. This weather had lower ground pressure field, lower ground and low-altitude wind speeds, and even calm wind sometimes. These were often accompanied by stronger radiation inversion or low-altitude temperature inversion that is difficult to eliminate. The structure of lower atmosphere layer was stable and might contribute to higher pollution.

3.3. Vehicle-mounted monitoring of pollutant discharge in some representative cities and key industrial parks

Quantitative modeling was done in some cities to evaluate the influence of key industries and industrial parks on ambient air quality, especially PM10. Vehicle-mounted DOAS system was used for monitoring industrial parks in both winter (Dec 2013) and summer (May-June, 2014) by way of acquiring data on spatial distribution of urban VOD of SO_2 and NO_2, emission characteristics of SO_2 and NO_2 in key industrial sources, as well as pollution transportation parameters under specific meteorological conditions.

At AY, the results obtained in winter and summer is similar. The concentration of SO_2 and NO_2 was caused mainly by the discharge from steelworks factories. However, observations during southeast/south winds in summer show that the concentration of SO_2, in the southwest-northwest section of the measurement route, was high. This indicates that sporadic small-scale industrial discharge occurred beyond southwest-northwest section (south and southwest of AY), which was not found during observations in winter. Judging from the relative geographic positions of steelworks and urban area at AY, as well as forward track model of wind field (Fig 7), it can be concluded that during northwest winds, discharge of AY steelworks had great influence on the air quality of urban area in AY.

Mean values per hour (MV/H), obtained from ground monitoring of SO_2 at monitoring point of environmental protection agency in AY from May 18 to May 20 are shown in Fig 8. From these results, it can be observed that ground concentration of SO_2 in urban area increased during northwest winds. At the same time, it is also found that during southeast winds, ground concentration of SO_2 in urban area was high at about 11a.m. on May 18, when the observation time was just the same as that of third auxiliary chart in Figure 9. It is concluded that, under southeast wind field, sporadic pollution in the southwest and south, beyond measurement area, had great influence on the air quality of main urban area in AY.

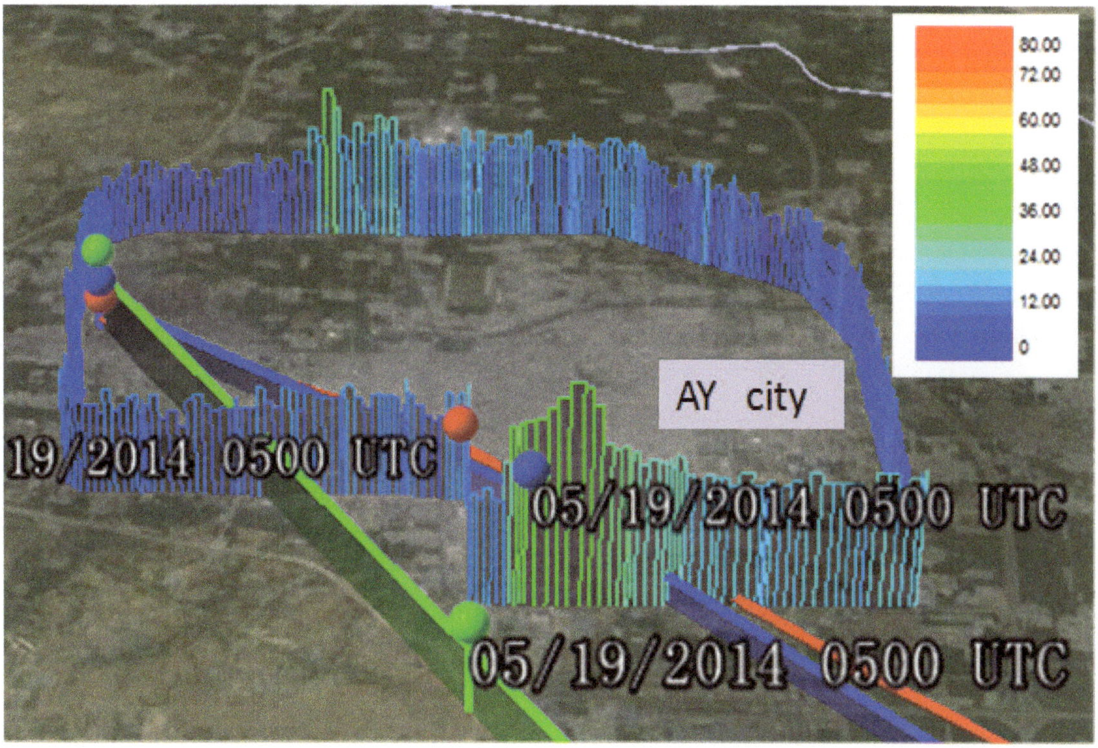

Figure 7. Spatial distribution of SO_2 and wind field track superposition at AY under northwest wind field

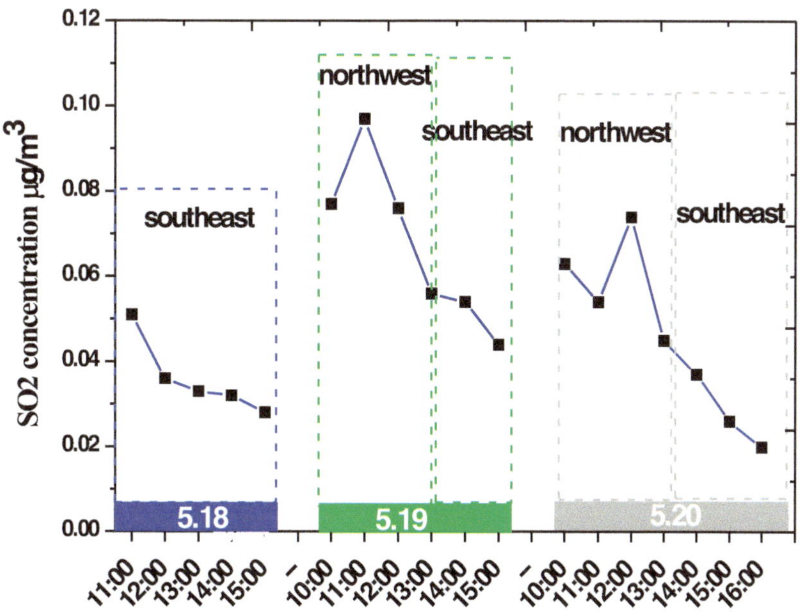

Figure 8. MV/H about ground concentration of SO_2 in monitoring point at AY

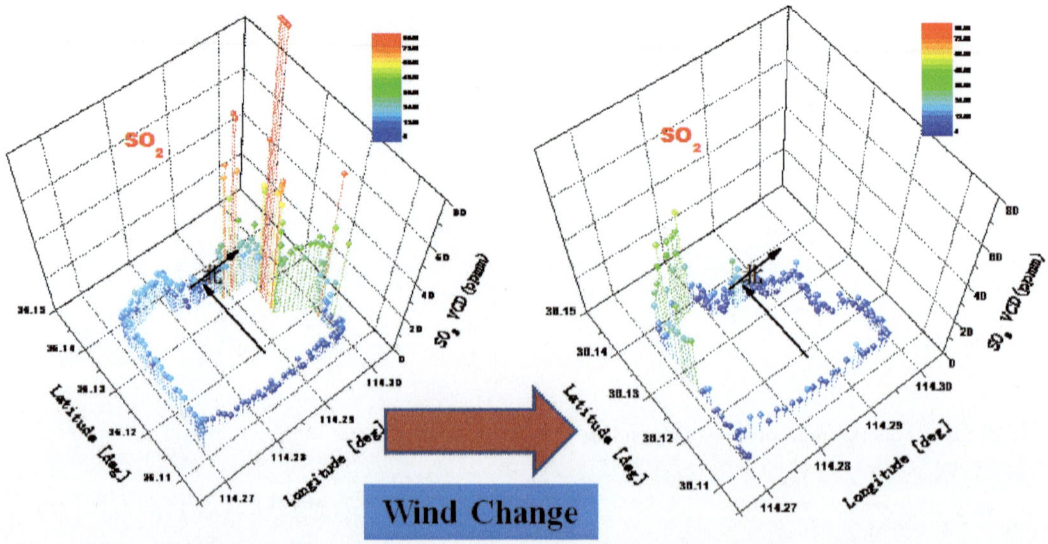

Figure 9. Spatial distribution changes of SO_2 during the wind filed transformation process at AY steelworks

3.4. Analysis of the typical haze process

Haze episode (Dec 12, 2013-Dec, 16, 2013)

From Dec 12, 2013 to Dec 16, 2013, a typical haze phenomenon was observed at AY, JZ and SMX, when the pollution was mainly local. Under the influence of regional weak wind field, the distribution range of local pollution was large and the duration long, but intermittent. Besides, the extinction coefficient of aerosol and concentration of NO_2 reflect a completely accumulative, rising, sustaining and stable process. Based on the distribution range and strength, two main phases were identified: generation phase of local pollution and pollution stabilization phase.

1. **Aerosols**

AY city: Since the afternoon of Dec 12, 2013, the pollutants gradually rose to 1.9 km and then began descending on Dec 13 (Fig 10). In the afternoon of Dec 14, it again started rising, but this time the height was lower than that on Dec 13. During Dec 15-16, near-ground pollution intensified and lasted for a long time, with the maximum value of extinction coefficient reaching 0.87km-1 (see Fig 11).

a. Formative period of haze (Dec 12-14)

b. Stable formation of pollution (Dec 15-16)

JZ city: The height of pollutants started rising from the afternoon of Dec 12 and the average of extinction coefficient was higher than that at AY (Fig 12). The height of daytime pollutants dropped on Dec 13, but it began rising again in the morning of Dec 14; near-ground pollution continued on Dec 15-16 (Fig 13).

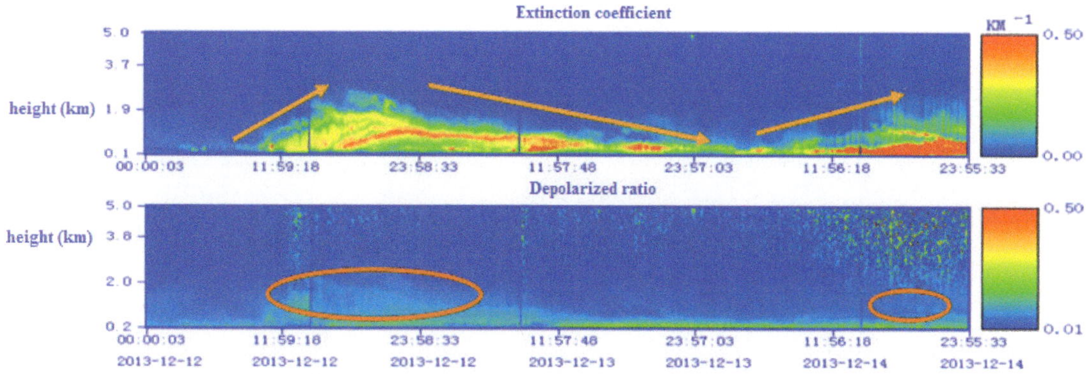

Figure 10. Spatial distributions of aerosols from Dec 12 to Dec 14 at AY, CCR

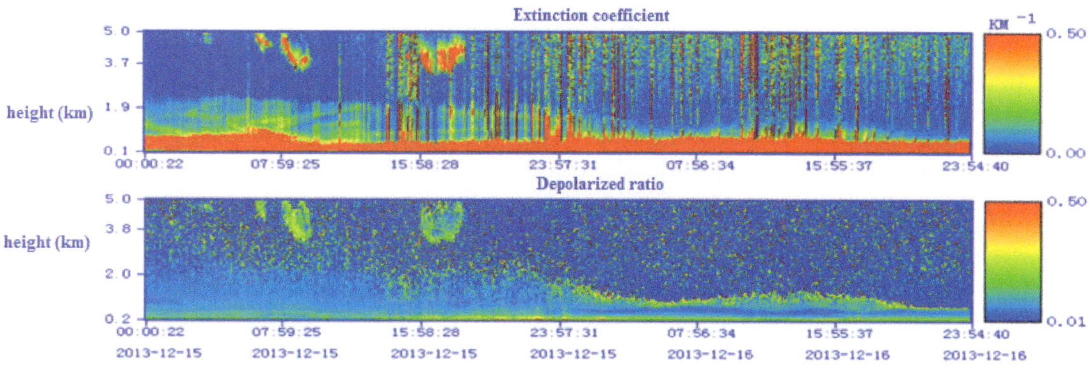

Figure 11. Spatial distributions of aerosols from Dec 15 to Dec 16 at AY, CCR

c. Formative period of haze (Dec 12-14)

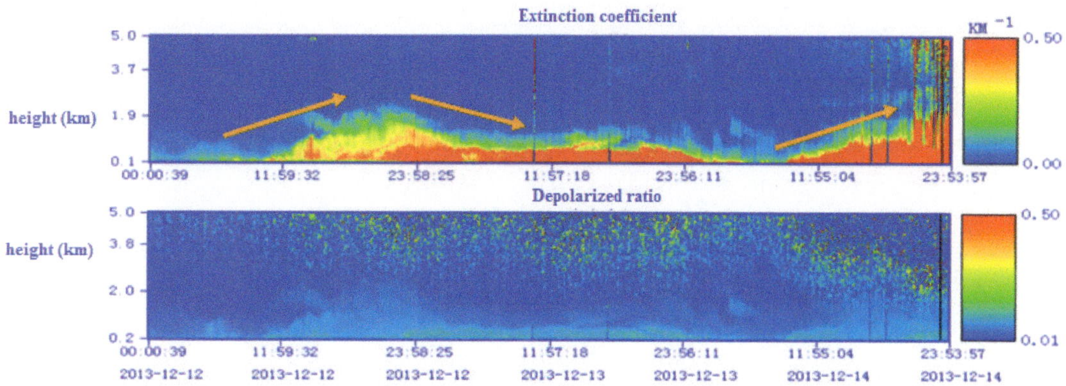

Figure 12. Spatial distributions of aerosols from Dec 12 to Dec14 at JZ, CCR

d. Stable formation of pollution (Dec 15-16)

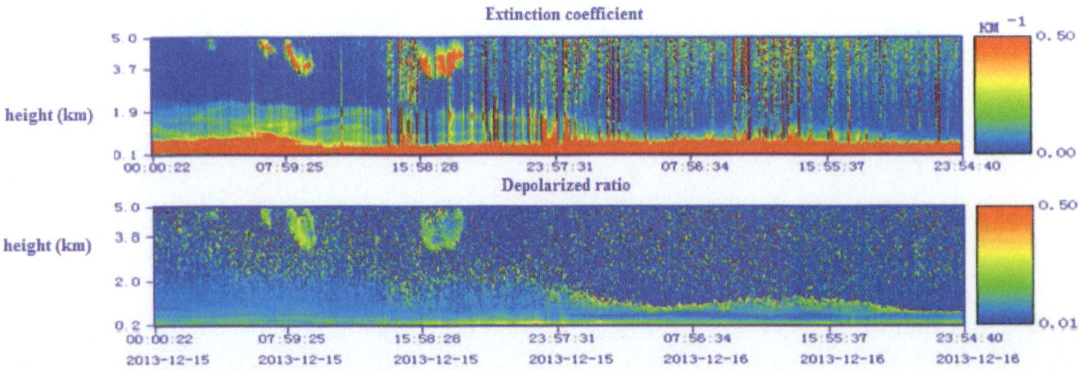

Figure 13. Spatial distributions of aerosols from Dec 15 to Dec16 at JZ, CCR

SMX city: Local pollution was noticed from the afternoon of Dec 12; the height of the pollutants rose gradually from the evening of Dec 13 and the average of extinction coefficient was 0.39km^{-1} (Fig 14). High concentration pollution was confined to the period from the evening of Dec 13 to evening of Dec 14. No obvious change of extinction coefficient profile is observed on Dec 15-16; the height of the pollutants was below 1.2km (Fig 15).

e. Formative period of local pollution (Dec 12-14)

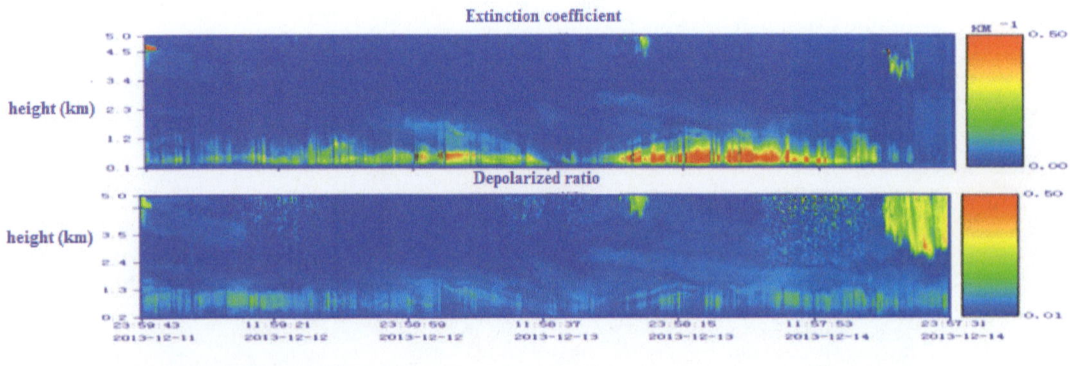

Figure 14. Spatial distributions of aerosols from Dec 12 to Dec 14 at SMX, CCR

f. Stable formation of pollution (Dec 15-16)

2. Pollution gas

During the haze period from Dec 14 to Dec 20, the whole process of NO_2 accumulation and dispersion, as observed by MAX-DOAS, is depicted in Figs 16 & 17. The changes in pollution concentration are most evident at AY, where the concentrate started rising since Dec 12 and

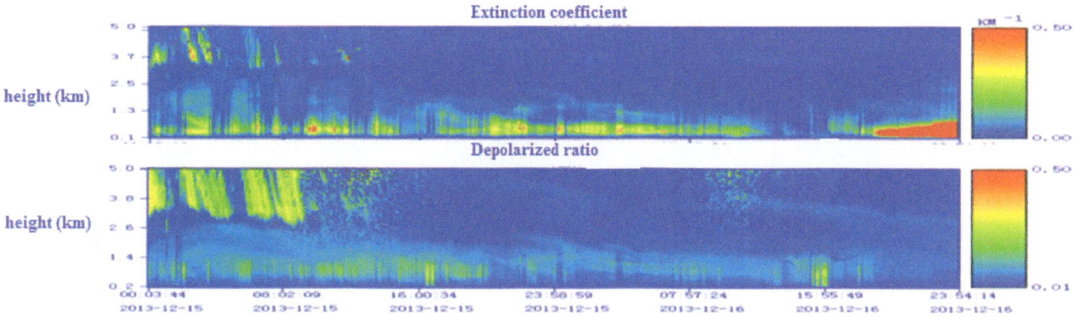

Figure 15. Spatial distributions of aerosols from Dec 15 to Dec16 at SMX, CCR

continued until Dec 14; thereafter, it remained at a high level. However, the increase at JZ was relatively slow. The concentration of both NO_2 and SMX NO_2 increased obviously on Dec 14 and again, and this time is more higher on December 18. On the contrary, the concentrations of NO_2 at AY and JZ began decreasing. The haze formation was caused mainly by internal and northern atmosphere transmission; besides, the obstruction to the airflow in the north militated against pollutant diffusion and instead favored accumulation. Air humidity, coupled with the convergence of cold and hot airflows, contributed to the process of haze formation.

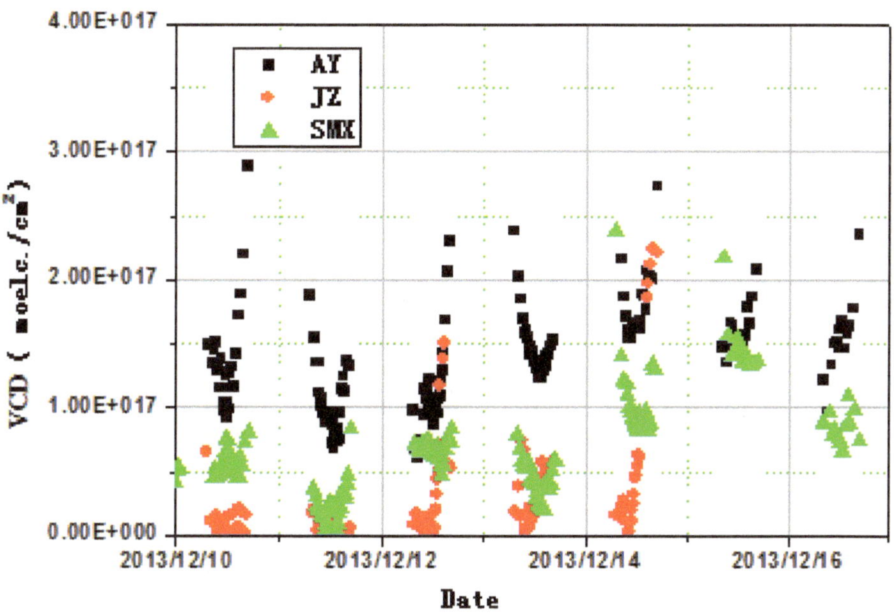

Figure 16. NO_2 VCD from Dec 10 to 16 in the observing sites

3. Meteorological conditions

On Dec 15, the average wind speed was 4.1 m/s and the temperature was1.8°C at the height of 1000m. The prevailing wind was northwest wind (Fig 18). The high-altitude west wind was

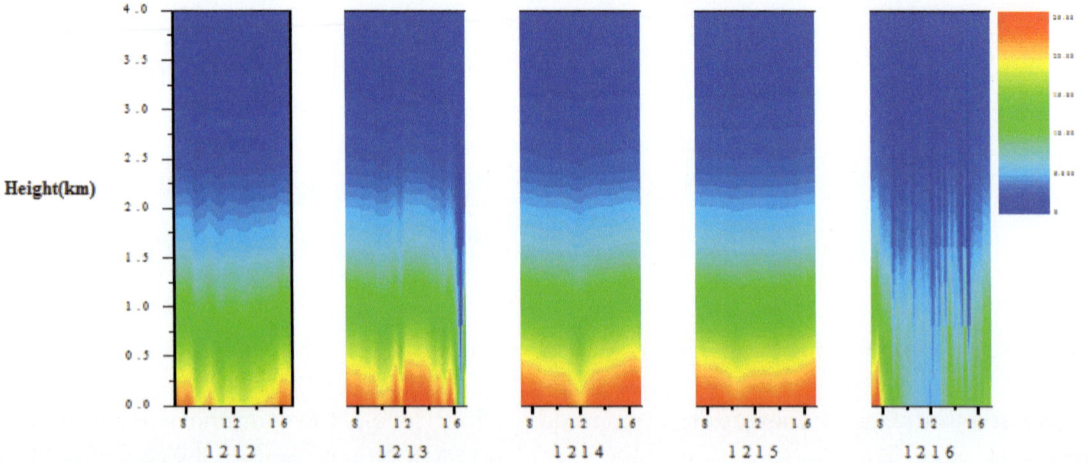

Figure 17. NO_2 profiles from Dec 12 to 16 at AY, CCR

affected by the downdraft in leeward slope terrain, leading to the convergence of the regional weak wind field pollutants, and finally to heavy pollution.

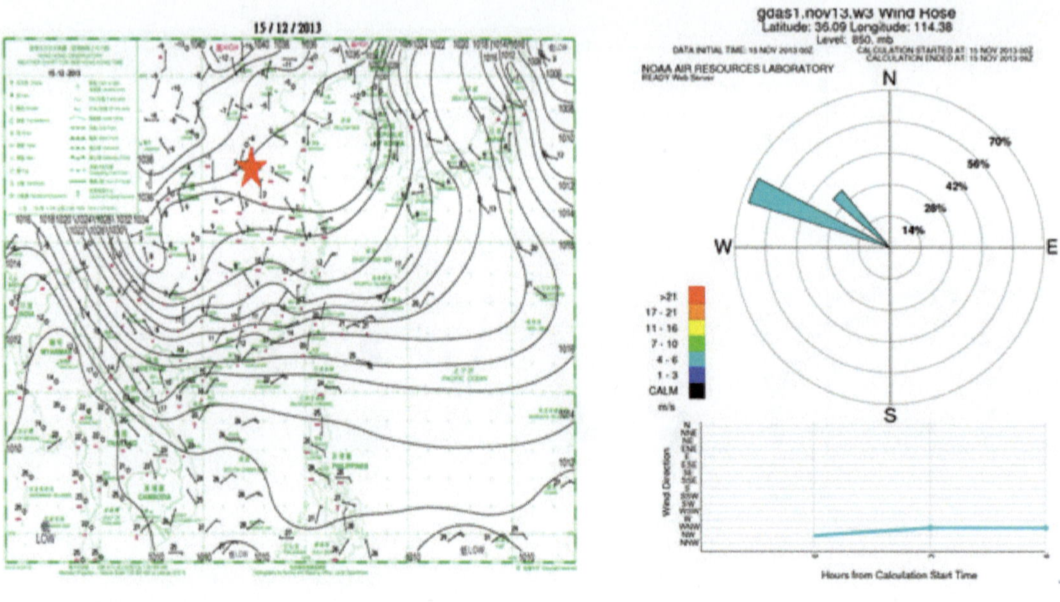

Figure 18. Meteorological conditions and wind rose diagram

From the HYSPLIT backward trajectory model in Fig 19 [29], the pollution at the three observation points was found to be mainly from the northwest. The main pollution source could be from burning of fossil fuels or bio-fuels, aided by the continental airstream. By noon on Nov 25, there was evident increase in pollutants at both AY and JZ. Subsequently, until the evening on Dec 26, at both the observation points, the pollutants decreased and then again

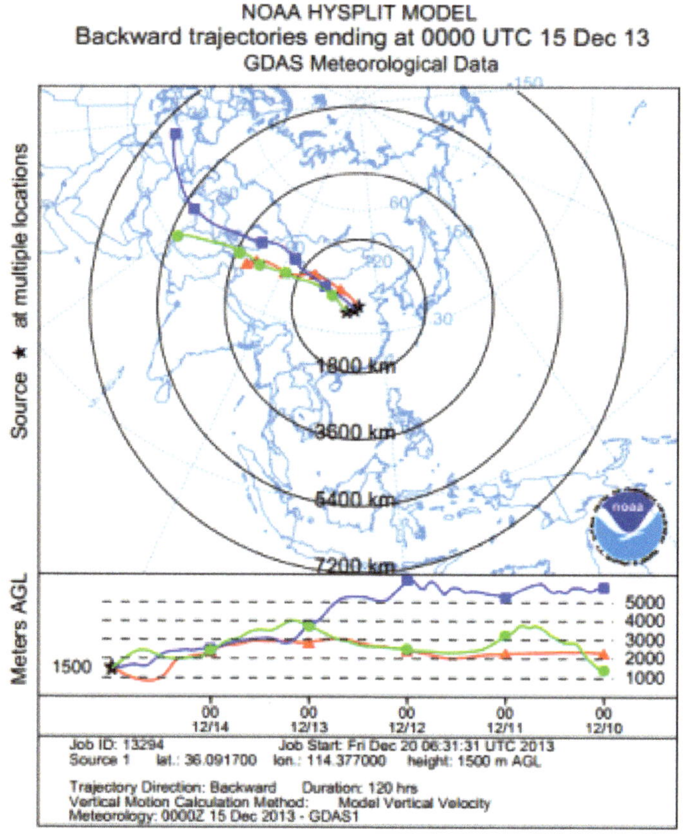

Figure 19. Backward trajectories from Hysplit

increased and stabilized. The overall aerosol extinction coefficient at AY was higher than that at JZ, and the peal values are 0.87km^{-1} and 0.51km^{-1}, respectively. The pollution layer was below 1.9 km, corresponding to the pollutant transmission layer in a low level jet (LLJ) [30] of stable or neutral boundary layer. For two days, it was observed that the height of the mixing layer grew linearly over time. When the height of the mixing layer grows to the height of the pollutant transmission belt in the nocturnal boundary layer, the vertical mixing effect would result in rapid concentration of pollutants on the ground. Therefore, there was increase in aerosol values, as well as in local pollutants discharge during the days; the downward transmission from the higher layers was also an important factor. Compared with AY and JZ, the air at SMX was cleaner and the height of pollutants was lower at 1.2 km below, but by the evening of June13 the pollution became serious.

4. Satellite

From the satellite results in Fig 20, it can be observed that high NO_2 VCD values existed during the haze period. From the daily variations in NO_2 VCD values between Nov and Dec at AY, JZ and SMX (seen in Fig 21), it can be seen that the results at AY clearly indicate a relation between NO_2 VCD values and haze. For example, at AY, the daily average value of NO_2 VCD

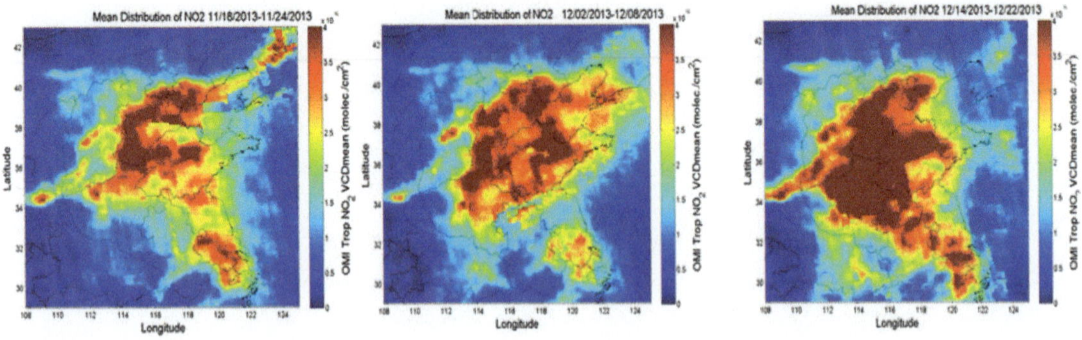

Figure 20. NO_2 VCD distribution result chart from Nov 11 to Dec 20 in CCR during haze period

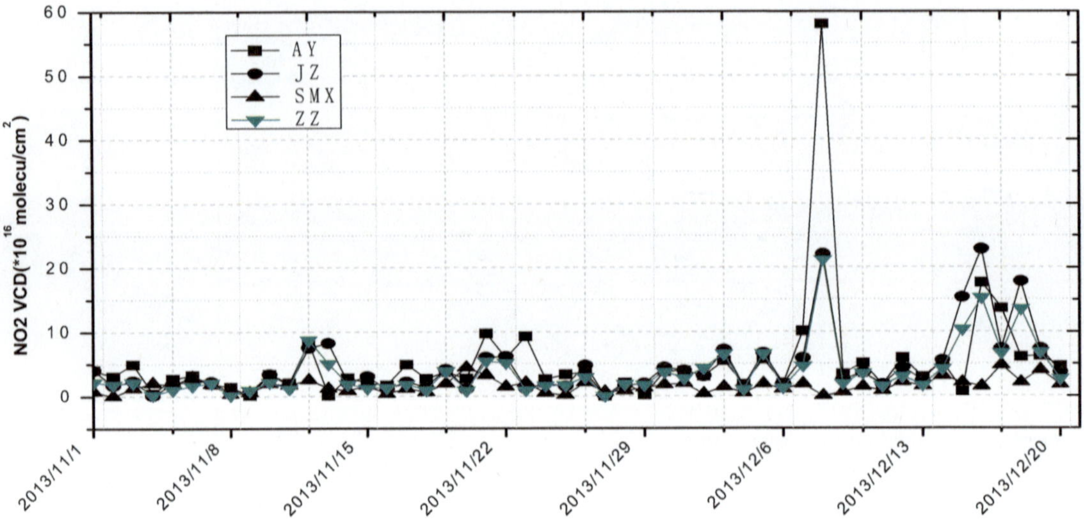

Figure 21. NO_2 VCD change chart detected by OMI satellite from Nov 1 to Dec 20

during Nov 1-20 was 3.32×10^{16}molec./cm^2. For the first haze period during Nov 21-23, the daily average value was 8.37×10^{16} molecule/cm^2, which was 2.5 times more than the normal average value. For the second haze period on Dec 8, it was 5.8×10^{17} molecule/cm^2, which was the highest and 18 times more than the normal average value. For the third haze period during Dec 16-17, it was 1.57×10^{17} molecule/cm^2, which was 4.7 times more than the average value. From the values of NO_2 during these three haze periods, which were multiple times more than the average value, it clearly follows that there exists certain relation between NO_2 and haze, and that high NO_2 VCD values contribute to hazy weather.

3.5. Typical straw burning process analysis (June 12-16, 2014)

1. **Aerosols**

Wheat harvesting from June 12 to June 16 in early summer of 2014, straw burning contributed to local pollution at all the three observation points in AY, JZ and SMX (Figs 22 & 23). At all these points, pollution was at its peak during June 14-15. And the biomass burning, such as straw burning, which releases particulate matter of small grain size, is confirmed by the depolarization ratio values from lidar observations.

June 13, 2014-June 14, 2014 (initial phase)

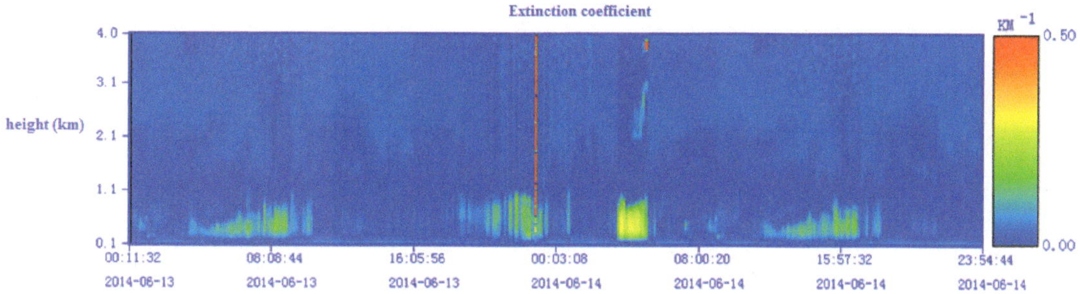

Figure 22. Spatial distribution of aerosol from June 13 to June 14, 2014 at AY, CCR

June 15, 2014-June 16, 2014 (serious pollution phase)

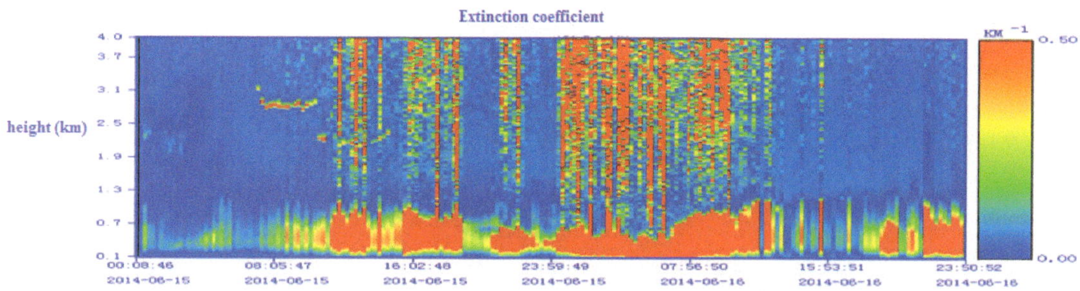

Figure 23. Spatial distribution of aerosol from June 15 to June16, 2014 at AY, CCR

During the straw burning period in CCR, the aerosol extinction coefficient fluctuated from 0.69 to 1.23 km^{-1}. The low depolarization peak value of 0.2 confirms the presence of fine particulates. Besides particulates, usually straw burning also releases various gaseous components such as CO_2, CO, NOx, VOC, sulfuric acid gas, hydrochloric acid gas, and alkaline gas ammonia. The retention time of ammonia in the air is short, because it quickly reacts with acidic gases or acidic particulate matter, and the resulting components can be precipitated on the particle surfaces of the particulate matter leading to the generation of new condensation nucleus. The aging process will increase the amount of particulate matter, which can lead to the increase in particulate emission factors. The burning of straw near CCR, together with surrounding pollution transmission, aging of particulate matter during long distance transmission, and the emission factors such as diameter aggravated local pollution.

2. Pollution gas

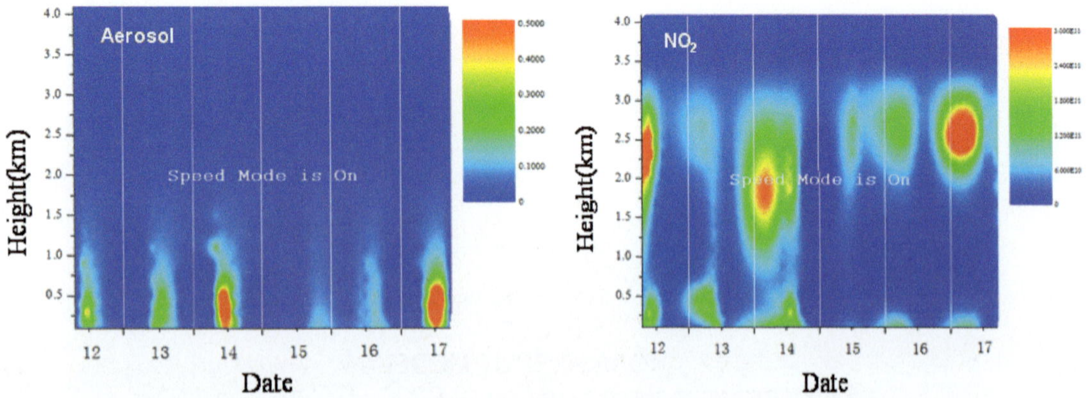

Figure 24. Aerosol and NO_2 profiles from June 15 to June16, 2014 at AY, CCR

NO_2 and SO_2 values were at low level for most of the time, but at some stages during the measurement period from the end of May to the middle of June, they were at high level. According to the results of aerosol and NO_2 for the period June 12 to 17 (see Fig 24), high aerosol extinction appeared at 300 m on June 12; it reached the peak on June 14 and diffused on June 15. It appeared on June 16 again and reached the highest level on June 17. NO_2 transmission process appeared at the height of 1.8 to 2.7 km on June 12; it weakened on June 13 although the transmission close to the ground was stronger. The transmission process appeared again at the height of 1.5 to 2.2 km on June 14, which reappeared, though as a weaker one, on June 15 and 16 at the height of 2.2 to 3 km. But the transmission remaining at the same height became stronger on June 17.

3. Fire spots and backward trajectory

To identify the sources of pollution transmission, local information was taken as the reference. But during the wheat harvesting season, the burning of straw on the surrounding farmlands was believed to be the main cause for high values of pollution. Therefore, the analysis was made by referring to the satellite fire spots diagram of Ministry of Environmental Protection, which is shown in Fig 25.

From these fire spots [31], it was inferred that multiple fire points existed around AY and JZ since June. However, by the middle of June they increased and many more fresh points appeared in the agricultural area of south CCR in the southeast. The backward trajectories, shown in Fig 26, indicate that concentrated burning of wheat straw brought about peripheral transmission.

From the foregoing results of aerosols and pollution gas, it emerges that the transportation brought by wheat straw burning manifested mainly as rapid increase of PM10, which is related to moisture absorption and expansion of particulate matter during transportation process. NO_2 and SO_2 were at their peaks on June 13 and 14 respectively; so, the amplification was not

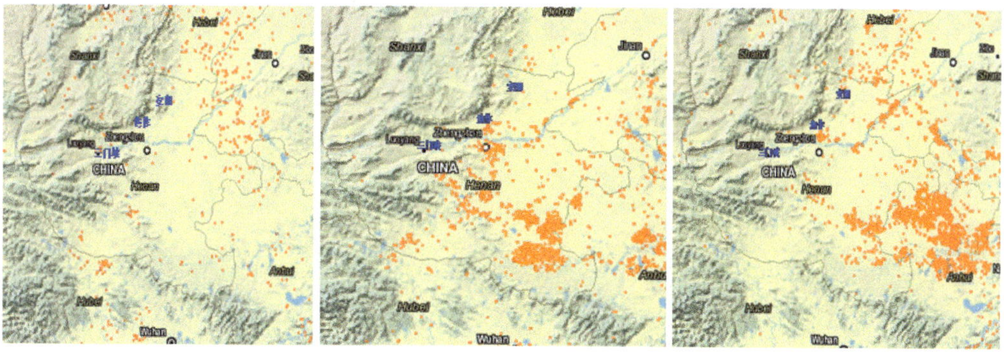

Figure 25. Fire spots from May 26 to June 1, June 2 to June 8, and June 9 to June 15

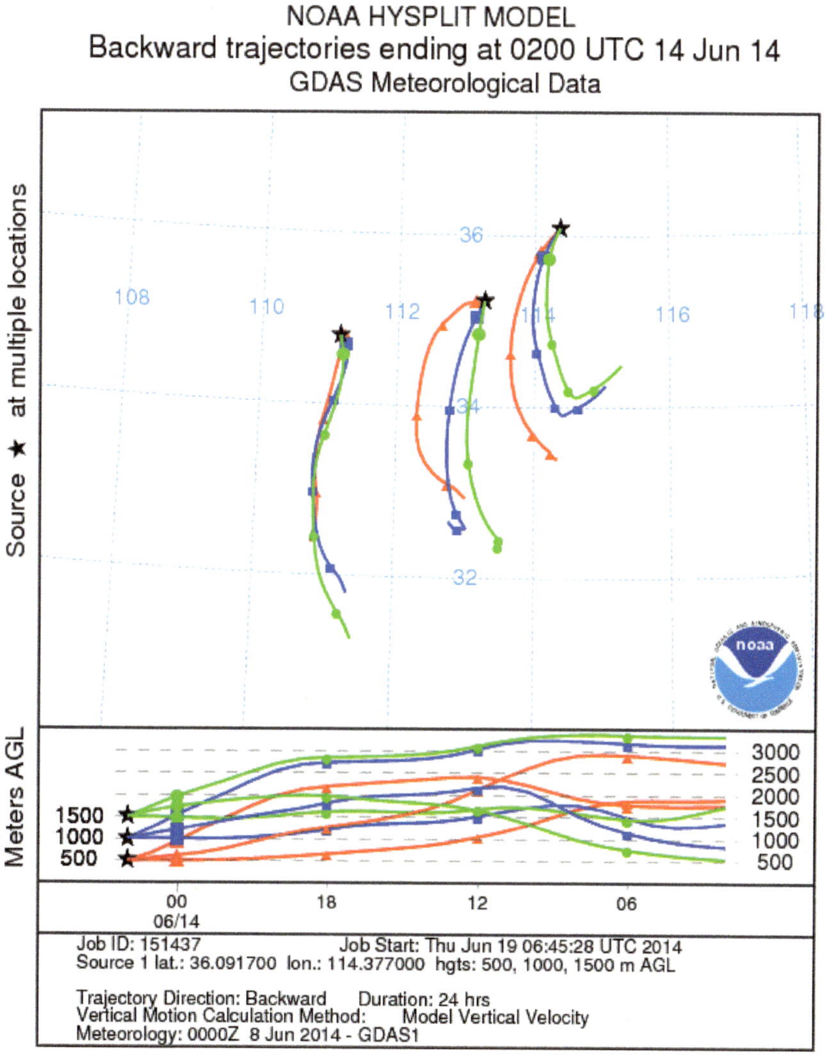

Figure 26. Backward trajectories from Hysplit

as obvious as that of PM10. For evaluating the influence of wheat straw burning on local air pollution, the values of NO_2 and SO_2 on June 11, before wheat straw burning, were taken as the background values. The NO_2 values obtained after wheat straw burning on June 13, and again the SO_2values obtained on June 14, were recorded and evaluated. From these evaluation results, it's concluded that from the end of May to the middle of June, concentration of gaseous pollutants (NO_2 and SO_2) and particulate matter increased obviously due to straw burning in CCR and surrounding cities.

4. Conclusions

CCR is an important industrial site in the mid- west region of China. Historical review of satellite data shows that the cities with relatively serious air pollution are confined mainly to the north because of their geographic location, weather conditions and industrial activity. Besides, the "accumulation zone" of pollutants exists in the region east of Tai-hang Mountains. Analysis of the data obtained from stereoscopic monitoring of the selected sites shows that the prevalence of pollution at the three sites, AY, JZ and SMX, was almost simultaneous. However, AY was characterized by local pollution (local source) and JZ by near-ground pollution, where the pollution concentration was also higher due to higher emission of industrial gases. Though local pollution was milder at SMX, it still can be easily aggravated by northwest sand dust transportation. Through statistical analysis of satellite and ground monitoring data, pollution was more serious during the 1st and 4th quarters (winter and spring), followed by the 2nd quarter and the 3rd quarter. During the heating period, the air pollution intensified obviously, because the urban energy consumption structure was still coal-oriented. During late May to mid-June, straw burning in the south CCR and surrounding provinces, such as Anhui and Jiangsu, contributed to the increase of aerosol, NO_2 and SO_2, in some cities of CCR.

As for the reasons for the complex process of haze formation, it was found that local pollution (local sources) was the main cause, besides the effect of superposition of pollution transportation in some time frames. Near-ground pollution at the three sites was caused mainly by local pollution (industrial emission, motor vehicle exhaust, coal combustion, etc.) and pollution transportation of some northern cities in the southeast direction. AY was more affected by local transportation and emissions from some industrial parks, and JZ by near-ground local pollution. Moreover, pollution concentration was higher at JZ than that at AY because of more industrial emissions. Local pollution at SMX was comparatively less obvious, but was lasting longer due to the unfavorable diffusion condition. In CCR, formation of ground weather situations of uniform-pressure field and North China low pressure, which are bad for pollutant diffusion, is easy. Such weather has lower ground pressure field, and smaller speeds of ground and low-altitude winds, which are sometimes calm. It is often accompanied with stronger radiation inversion or low-altitude temperature inversion that is difficult to eliminate. The structure of lower atmosphere layer is stable and might contribute to higher pollution.

So far many efforts were made from ground based and space borne observations to study aerosol distribution and properties, along with model simulations, but such integrated studies are still limited over the East Asian, especially in CCR region in China. Satellite observations provide time constrained observations of aerosol vertical distributions but ground based lidar can give information on the evolution of the distribution over a location. Thus the optical properties of aerosols or gaseous pollutants obtained from ground-based remote sensing technology, especially lidar and DOAS is being studied to determine the aerosol optical characteristics and source origin. In the future, the more integrated combination of lidar, satellite data, in-situ instruments, and sunphotometer data will help us to improve the characterization of a variety of aerosol events and get deeper insight into the mechanism of long-range transportation of pollution in China. Furthermore, the information of vertical distribution of haze layer during long-range transportation could permit us to make better assessment of the effect of aerosols on the radiative forcing and climate change.

Acknowledgements

This work was supported by the Funding from Henan Program (Researches on Atmospheric Haze pollution in Henan Province), Major State Basic Research Development Program of China (2014CB447900) and NSFC (41305126). The authors are also thankful to the NOAA Air Resources Laboratory (ARL) for providing the HYSPLIT transport and dispersion model used in this research.

Author details

Wenqing Liu, Zhenyi Chen*, Pinhua Xie, Jianguo Liu and Tianshu Zhang

*Address all correspondence to: zychen@aiofm.ac.cn

State Environmental Protection Key Laboratory of Optical Monitoring Technology, Ministry of Environmental Protection of the People's Republic of China, Anhui Institute of Optics & Fine Mechanics, Chinese Academy of Sciences, Hefei, Anhui, P.R. China

References

[1] Ramanathan, V., P. J. Crutzen,, J. T. Kiehl, and D. Rosenfeld, Atmosphere: aerosols, climate, and the hydrological cycle. *Science*.,2001, 294, 2119–2124.

[2] Ian G. C., Guand C. L. and Jami H., China's Growing Urban Health Inequalities: The Challenges Ahead, J. Management and Sustainability, 2013, 3(2):10-18

[3] Sun Y. L., Zhuang G. S., Wang Y., Han L. H., Huo G. J. and Dan M., The air-borne particulate pollution in Beijingconcentration, composition, distribution and sources. *Atmos. Environ.*, 2004,38(35): 5991–6004

[4] Tie, X., Brasseur G., Zhao C. S., Granier C., Massie S., Qin, Y., Wang P. C., Wang G. L., Yang P. C. and Richterg A., Chemical characterization of air pollution in Eastern China and the Eastern United States. *Atmos. Environ.*, 2006,40, 2607–2625

[5] Wu, D., Deng X. J., Bi X. Y., Li F., Tan H. B., Liao G. L., Study on the visibility reduction caused by atmospheric haze in Guangzhou area. *J. Tropical Meteorology,* 2006,13, 77–80

[6] Zhang, Y. H., Hu M., Zhong L. J., Wiedensoher A., Liu S. C., Andreae M. O. and Fan F. J., Regional integrated experiments on air quality over Pearl River Delta 2004 (PRIDE-PRD2004), *Atmos. Environ.*,2004,42,6157-6173

[7] Wang Y.S., Xin J. Y., Li Z. Q., Wang P. C., Wang S. G., Wen T. X., Sun Y., AOD and Ångström Parameters of Aerosols Observed by the Chinese Sun Hazemeter Network from August to December 2004, *Environ. Sci.*, 2006,27(9)

[8] Tesche, M., Ansmann A., Müller D., Althausen D. and Engelman R., Particle backscatter, extinction, and lidar ratio profiling with Raman lidar in south and north China, *Appl. Opt.*, 2007,46(25):6302-8

[9] Xiao, R., Takegawa N., Zheng M., Kondo Y., Miyazaki Y., Miyakawa T., et al.,Characterization and source apportionment of submicron aerosol with aerosol mass spectrometer during the PRIDE-PRD 2006 campaign, *Atmos. Chem. Phys.*, 2001,11, 6911-6929

[10] Li, Z., et al., Preface to special section on East Asian Studies of Tropospheric Aerosols: An International Regional Experiment (EASTAIRE), *J. Geophys. Res.*, 2007,112, D22S00, doi:10.1029/2007JD008853.

[11] Li, Z., et al., East Asian Studies of Tropospheric Aerosols and their Impact on Regional Climate (EAST-AIRC): An overview, *J. Geophys. Res.*, 2011,116, D00K34, doi: 10.1029/2010JD015257

[12] Atsushi, S., Nobuo S., Ichiro K. A., Itsushi U., Toshiyuki M., Naoki K.,et al., Continuous observations of Asian dust and other aerosols by polarization lidars in China and Japan during ACE-Asia, *J. Geophys. Res.*,2004 109, D19S17

[13] He, Q. S., Li C. C., Mao J. T. and Li P. R., A study on the aerosol extinction-to-backscatter ratio with combination of micro-pulse LIDAR and MODIS over Hong Kong, *Atmos. Chem. Phys.*, 2006, 3243–3256

[14] Müller, D., Tesche M., Eichler H., Engelmann R., Althausen D., Ansmann A., et al,Strong particle light absorption over the Pearl River Delta (south China) and Beijing (north China) determined from combined Raman lidar and Sunphotometer observations, *Geophys. Res. Lett.*, 2006, 29 (17)

[15] Mattias J., Bo G., Tong Y., Lin T., Chen D.L., Li H.J., et al., Quantification of total emission of air pollutants from Beijing using mobile mini-DOAS, 2008,*Atmos. Environ.*,42(29), 6926–6933

[16] Meng X.L., Fan X.Y., Meng X., Zheng J.P., Mason R.,Submicron-sized fluid inclusions and distribution of hydrous components in jadeite, quartz and symplectite-forming minerals from UHP jadeite–quartzite in the Dabie Mountains, China: TEM and FTIR investigation, 2009,*Applied Geochemistry*,24(4), 517–526

[17] Cheng, Y., Lee S. C., Ho K. F., Wang Y. Q., Cao J. J., Chow J. C. and Watson J. G., Black carbon measurement in a coastal area of south China, *J. Geophys. Res. Atmos.*, 2006, 111, D12310, doi: 10.1029/2005JD006663

[18] Franke, K., A. Ansmann, D. Muller, D. Althausen, C. Venkataraman, M. S. Reddy, F. Wagner and R. Scheele (2003),Optical properties of the Indo-Asian haze layer over the tropical Indian Ocean, J. Geophys. Res., 108, D2, 4059, doi: 10.109/2002JD002473

[19] Klett, J. D., Stable analytical solution for processing lidar returns, *Appl. Opt.*, 1981, 20, 211–220,

[20] Frederick G. Fernald. Analysis of atmospheric lidar observations: some comments, *Appl. Opt.*, 1984, 23(5), 211–220

[21] Xu J., Xie P.H., Si F.Q.,Li A.,et al., Comparison of NO2 slant columns between two ground-based Max-DOAS, *Spectroscopy and Spectral Analysis*, 2012,32(2),558-564

[22] Wang Y., Li A., Xie P.H., Chen H., Xu J.,Wu F. C., Liu J.G., Liu W.Q., Retrieving vertical profile of aerosol extinction by multi-axis differential optical absorption spectroscopy, *Acta Phys. Sin.*,2013, 62(18), 180705

[23] Wu J.G. Modern Fourier Transform Infrared Spectroscopy and Its Application, Beijing: Scientifical and Technical Documents Press,1994,1, 492-505

[24] Manne K, Ph. D. Dissertation, Sweden: Chalmers University of Technology.2005

[25] Fransson, K and Mellqvist, J., Measurements of VOCs at Refineries Using the Solar Occultation Flux Technique, Report to Preem Environmental foundation, Göteborg, 2002

[26] Wang T., Wang P.C., Yu H., Zhang X.Y., Zhou B.,Si F.Q., et al., Intercomparison of slant column measurements of NO2 by ground-based MAX-DOAS, *Acta Phys. Sin.* 2013, 62(5), 054206

[27] Wu F.C., Xie P.H., Li A.,Xu J., Shi P., Qin M.,Wang M.H., et al., Studies on the Remote Measurement of the Distribution ion of City Gaseous Pollutant by Mobile Passive Differential Optical Absorption Spectroscopy, *Spectroscopy and Spectral Analysis*, 2011,31(3),583-588

[28] Witte J.C., Duncan B.N., Douglassb A.R., Kurosuc T.P., Chancec K., Retscherd C.,The unique OMI HCHO/NO2 feature during the 2008 Beijing Olympics: Implications for ozone production sensitivity, *Atmos. Environ.*,2011, 45(18), 3103–3111

[29] Draxler, R. R., and Rolph G. D., HYSPLIT (Hybrid Single-Particle Lagrangian Integrated Trajectory) Model access via NOAA ARL READY, (2003) web site (http://www.arl.noaa.gov/ready/hysplit4.html), NOAA Air Resources Laboratory, Silver Spring, MD.

[30] Hu X.M., Klein P.M., Xue M., Zhang F.Q., Doughty D.C., Forkel R., Joseph E, Fuentes J.D., Impact of the vertical mixing induced by low-level jets on boundary layer ozone concentration, *Atmos. Environ.*, 2013,70,

[31] http://hjj.mep.gov.cn/stjc/

5

Biological Contamination of Air in Indoor Spaces

Anca Maria Moldoveanu

Additional information is available at the end of the chapter

1. Introduction

The indoor environment is a very complex environment. Pollutants are numerous in both outdoor and indoor air. While the same pollutants (biological, chemical, or radioactive) can be identified in indoor and outdoor spaces, their concentrations can be very different. More specifically, extremely high concentrations can accumulate indoors. Many bacteria that belong to the normal microflora of the human skin are continuously emitted into the immediate surroundings of humans, and accumulate indoors.

The pollution of the indoor spaces is an important element that can affect human health. People spend around 90% or more of their time of a day indoors. Indoor spaces include the inside of houses, schools, university rooms, social buildings like shops, cars, planes, and workplaces. The outdoor and indoor environments are linked. The two elements cannot be separated. Biological agents are defined as agents or microfragments from plant or animal matter or from microorganisms.

Elements identified in the air include bacteria, either in vegetative status or spores, fungi, yeasts, microbial toxins and secondary metabolites like bacterial endotoxin, peptidoglycans or fungal β(1,3)-glucans, volatile organic compounds, pollens, pet and insect allergens, other allergens, viruses, protozoa, etc. [1-3].

Many bacteria, fungi, viruses, and protozoa may be infectious to humans and they may cause specific infectious diseases. Infectious agents are usually received from other humans, from animals (these infections are zoonoses), or from environmental sources such as contaminated water, soil, etc.

Biological contamination in indoor air may have sources like outdoor air, human body, bacteria that are growing indoors, and pets.

2. Biological contamination of indoor air

2.1. Bacteria in the indoor environments

The most common building-associated bacteria are saprophytic bacteria of the normal human skin, mouth, and nose that are emitted into the indoor air and bacteria originating from outdoors air [1-3]. Other common bacteria of indoor environments include heterotrophic bacteria that grow in the water reservoirs or moist sites of the building, such as bathroom sinks. Specifically, legionellae and nontuberculous environmental mycobacteria develop in biofilms of water pipelines or in water reservoirs of cooling systems. Actinobacteria (especially Streptomycetes), *Bacillus* species and various other bacteria grow in moist building materials together with fungi. Elements from bacterial structures released in air include bacterial cells, bacterial spores, peptidoglycans, microbial volatile organic compounds, exotoxins, and other bacteria growing metabolites [1].

Humans are an important source of indoor bacteria. The upmost layer of the normal human skin is continuously renewed, and skin scales containing bacteria are shed into the environment [1]. Bacteria in the respiratory airways are eliminated through Pflügge droplets while talking, coughing, or sneezing. The level of air contamination is dependent on the number of persons inside a room and the efficiency of the ventilation system (natural or artificial ventilation). Bacteria that can be identified in indoor air are micrococci, staphylococci, streptococci, and corynebacteria [1].

The presence of pets in indoor spaces can represent an important source of air contamination and can be linked with the level of indoor endotoxins. The presence of dogs and cats can be the main predictors of endotoxin levels in house dust [1, 4-7]. Other predictors are the presence of vermin, such as mice, and infrequent cleaning, which indicates poor hygienic conditions in the home [1]. Storage of organic household waste indoors also increases bacterial contamination in the indoor environment [1].

Bacteria actively growing or accumulating in the indoor environment may cause health effects and may require specific actions to control growth and prevent the exposures.

Bacterial growth can be found wherever standing water is present and Gram-negative bacteria can be identified in this environment. The presence of bacteria, either as viable bacteria or bacterial spores, mycotoxins, chemical markers like β-glucans and volatile organic compounds, and endotoxins, represents an indication of high humidity in the indoor environment [1].

2.2. Fungi in indoor environments

Fungi are present in both outdoor and indoor air. The levels found in the outdoor air can range from less than 100 spores/m^3 up to100 000 spores/m^3. Normally, the fungi levels are lower in indoor spaces compared with outdoor spaces. If the conditions that permit the fungal growth are present, for example, in damp houses, the levels of fungi in the indoor air can reach very high concentrations. Fungi level measurements are expressed in colony-forming units (CFUs) per cubic meter of air.

The levels of fungi and the type of fungi identified in indoor air depend on the season, construction features, age, and use of the building as well as the ventilation rate.

The most common fungal genera occurring in indoor environments are *Penicillium, Aspergillus, Cladosporium, Alternaria,* and yeasts [1-3]. These genera are also the most frequently occurring fungi in outdoor air.

Spores of fungi are present everywhere, and they are able to germinate wherever there is water available and the ranges between 0.80 and 0.98. Fungi need carbohydrates, proteins, and lipid to develop. They can find all these elements in house dust, construction materials like wallpaper or textiles, paint, glue, wood, paper and books, stored food, or deposit of cooking oil. Fungi can grow on inert materials like ceramic tiles [2].

2.3. Allergens in indoor environment

Allergens are a group of agents that may cause a specific IgE-mediated reaction in humans. Fungal allergens can trigger type I allergic reactions and IgE sensitization to fungal species like *Alternaria, Penicillium, Aspergillus,* and *Cladosporium* spp. All these types of fungi species can induce allergic respiratory diseases, especially acute asthma attacks. Fungal allergens can also trigger type III-IgG-mediated allergic reactions (*Penicillium* and *Aspergillus)* and even type IV allergic reactions. This is the premise of the onset of hypersensitivity pneumonitis [2].

Allergens may also be present in indoor air due to house dust mites. The mite species that produce allergens are *Dermatophagoides pteronyssinus* (major allergens—Der p I and Der p II) and *Dermatophagoides farina* (major allergen—Der f I).

2.4. Endotoxins

These are toxins integrated in the outer membrane of Gram-negative bacteria, which are composed of proteins, lipids, and lipopolysaccharides [2, 4-8]. The bacterial cell lysis determines the release of endotoxins. In the air, endotoxins are bound to coarse particles. Exposure to different levels in air has different results: if the exposure is to high levels, respiratory symptoms or nonallergic asthma can develop, while if the exposure is at low and moderate levels, it is possible to have a protective action against allergies and asthma [2, 4-8].

2.5. (1 → 3)- β -D-glucans

It is a structural element of the cell wall of fungi and some bacteria, and it can be found in house dust. (1 → 3)-β-D-glucans do not have allergic properties. However, they do have immune-modulating properties. Exposure to (1 → 3)-β-D-glucans in home environment can induce respiratory diseases [2, 9, 10] and may increase peak flow variability in asthmatic children.

2.6. Mycotoxins

These are toxins (secondary metabolites) produced by fungi that interfere with RNA synthesis and can cause DNA damage. Some mycotoxins like aflatoxins produced by *Aspergillus flavus* and *Aspergillus parasiticus* induce carcinogenic effects [2, 11]. Indoor exposure to mycotoxins produced by *Stachybotrys chartarum* can be associated with acute pulmonary hemorrhage.

2.7. Fungi and Volatile Organic Compounds (VOC) in indoor air

Fungi can produce around 200 volatile organic compounds. The identification of the VOCs in indoor air shows that there is a fungal growth present, even if the quantitative measurements are negative [11]. One of the most important substances that can be identified in indoor air is formaldehyde that can induce cancer.

2.8. Viruses

Indoor air humidity can induce a longer survival time of respiratory viruses and increased risk of respiratory infection and allergic diseases.

3. Exposure evaluation of biological contamination of indoor air

Indoor air levels of bacterial concentrations of <1, 000 CFU/m^3 can be regarded as "low, " and concentrations of >5, 000 CFU/m^3 can be regarded as "high" [1]. These numbers reflect mainly the crowdedness of the room and the efficiency of the ventilation and thus are a measure of the hygienic quality of the indoor air.

Characterizing the bacterial flora in the sample may give additional information about the sources of the bacteria present. If Gram-positive cocci and corynebacteria, the common bacteria of the human skin, are the dominating groups, this indicates human sources [1].

The presence of Gram-negative bacteria shows an active bacterial growth due to the presence of water indoors.

Gram-positive bacteria with exo- and endospores like *Streptomyces* and *Bacillus* can grow on moist building materials. Their spores are very resistant and can survive even if the air humidity is low.

The presence of fecal bacteria like the group of *Enterobacteriaceae* is considered heavy contamination of indoor air, coming probably from a sewage leakage. The presence of actinomycetes shows the growth of microbes in the building material.

The presence of two groups of bacteria, that is, *Legionella* and actinomycetes, has a special importance in indoor air.

Legionella spp. are Gram-negative environmental bacteria that have found a remarkable ecological niche in man-made water systems, especially in those serving air conditioning systems for heat transfer [1].

4. Health effects induced by indoor exposure to biological contaminants

Indoor air contamination can generate a multitude of effects on human health. The effects are especially on the respiratory system. Allergic reactions can be induced by the exposure to different biological agents that are identified in indoor air. Also, there is evidence to suggest that exposure to biological pollutants in indoor air can be associated with symptoms with neurological basis like headaches, fatigue, and forgetfulness.

Some of the effects on the respiratory system related to the biological contamination of indoor air are due to the inflammation of the airways. This can appear in both allergic (with the implication of IgE and IgG antibodies) and nonallergic respiratory diseases. The separation between those two entities is important in the case of asthma and rhinitis.

4.1. Rhinitis

Rhinitis is induced by the inflammation of the mucosa of the upper respiratory tract. Symptoms of rhinitis are nasal congestion, rhinorrhea, sneezing, and conjunctivitis. Rhinitis can be of allergic or nonallergic cause [1, 11].

4.2. Asthma

Allergic asthma involves IgE-mediated sensitivity and eosinophils in presence of an atopic underlying reaction of the respiratory airways [1, 12-18]. Among adults, there is also an increasing prevalence of asthma. People with moderate to severe asthma will react adversely toward biological and other contaminants in the indoor environment due to a nonspecific bronchial hyperresponsiveness. That is an exaggerated bronhoconstrictor response to numerous numbers of stimuli. The severity of asthma and the need of medication in house dust mite allergic patients is closely related to the amount of endotoxin in house dust in a dose-dependent manner [1].

Pollen allergens interact with other sources of fine particles in polluted air, which concentrate the allergens and trigger asthma attacks. As for the interaction between biological and nonbiological agents, airborne particles such as diesel exhaust particles have an adjuvant activity for IgE antibody and cytokine production [1]. The intranasal instillation of ragweed allergens with diesel exhaust particles in allergic patients significantly increased allergen-specific IgE production and enhanced Th2-type cytokine mRNA expression compared to allergic patients instilled with ragweed alone [1].

Both allergens and viruses have been known to worsen asthma symptoms. A study in the UK demonstrated that allergen exposure and viral infections act synergistically to trigger exacerbations in adults [1].

4.3. Chronic bronchitis

Chronic bronchitis is defined as cough with phlegm (productive cough) for at least 3 months per year for 2 years or more [1, 17-19]. Chronic bronchitis may be present with or without airways obstruction. The underlying mechanism is an increase in mucus secreting glands and alterations in the characteristics of the mucus itself [1]. Airway inflammation is present, and an increased number of neutrophils can be observed [1].

Subjects with chronic bronchitis may also have symptoms such as dyspnea, chest tightness, and wheezing, and there is thus a considerable overlap in symptoms between allergic asthma, nonallergic asthma, and bronchitis [1]. Biological contamination through the presence of endotoxin and β-glucan can be incriminated in the pathogenesis of bronchitis [1].

4.4. Chronic airflow obstruction

Several studies have shown associations between chronic airflow obstructions as measured by spirometry (particularly FEV1) and organic dust exposures in industrial and agricultural occupational populations [1, 19]. Endotoxin exposure may play an important role.

4.5. Hypersensitivity Pneumonitis (HP) or Extrinsic Allergic Alveolitis (EAA) [1, 2, 20]

The disease, also known as "farmers lung" or "pigeon breeder's lung, " is one of the most severe diseases induced by fungi and thermophilic actinomycetes (spore-forming bacteria) exposure. The clinical aspect of HP is represented by severe pulmonary disease with pneumonia-like fever, cough, tightness of chest, shortness of breath, inflammation of pulmonary parenchyma, and development of pulmonary granulomas. HP is very similar to organic dust toxic syndrome (ODTS).

4.6. Organic Dust Toxic Syndrome (ODTS)

The disease is similar to influenza and is characterized by symptoms like acute fever, dry cough, dyspnea, chest tightness, headache, shivering, malaise, fatigue, and muscle and joint pains. ODTS is a nonallergic neutrophil-mediated airway inflammation [1].

4.7. Atopic allergic dermatitis

This disease can be induced by exposure to the biological indoor air pollutants.

4.8. Sick building syndrome

In the 1988s, the concept of "sick building syndrome" (SBS) was established by the WHO [1, 3, 21, 22]. The label sick building syndrome (SBS) can be applied when the inhabitants or occupants of different indoor spaces develop acute health problems or discomfort inside the rooms. The exposed people have the following symptoms: eye, nose, and throat irritation and dryness; itching and dryness of skin and eyes; dry cough; nausea; vomiting; headache; difficulty in concentration; fatigue; etc. The symptoms disappear after cessation of the exposure.

5. Aim and purpose of the study

The aim of the study was to assess indoor air contamination from the point of view of bacteriological and fungal contamination of different indoor spaces like houses, schools, offices, archives, and spaces with commercial destinations [23-25].

6. Materials and methodology of investigation

The methodology of the study implies the investigation of the levels of indoor air contamination with fungi and bacteria. The destination of the investigated indoor spaces was very diverse. The levels of bacteriological contamination were assessed in indoor spaces designated as living spaces, social use spaces like shops, and work and education spaces like schools, offices, and archives.

Quantitative and qualitative assessments were made to evaluate the levels of air contamination. Passive and active methods were used to sample the air from the investigated sites. The results were compared to the enforced recommendations in Romania, at this moment.

6.1. Investigated places

6.1.1. Houses in Bucharest City

There were 10 investigated apartments, with different numbers of room situated at different floors and with different geographical orientations and different characteristics.

Type of investigated apartments:

a. One apartment with one room, one inhabitant, nonsmoker

b. One apartment with one room, two inhabitants, one smoking inhabitant

c. Apartment with two separates rooms, two nonsmoking inhabitants

d. Apartment with two separates rooms, three smoking inhabitants

e. Apartment with two rooms, with five nonsmoking inhabitants

f. Apartment with two separates rooms, four inhabitants, one smoking person

g. Apartment with three separates rooms, three nonsmoking inhabitants

h. Apartment with three separates rooms, five inhabitants, one smoking person

i. Apartment with three rooms, five nonsmoking inhabitants

j. Apartment with three rooms, with four inhabitants, one smoking person

For an accurate assessment of the biological contamination of indoor air, the measurements were made three times in a year, in the winter, in the summer, and in the autumn, 3 days consecutively, in the kitchen and in the living room of each investigated apartment.

6.1.2. Schools

There were two investigated schools, one in Bucharest and one in Giurgiu City.

In the school in Bucharest, the investigations were made

- in an empty classroom, before 8:00 a.m.
- during the school day (at 9.30 a.m.), in a classroom with 23 children
- in a classroom with 36 children and 1 teacher
- in a classroom with 23 children and 1 teacher
- in a classroom (at 11:00 a.m.) with 24 children and 1 teacher
- in the school library (at 10:00 a.m.) with 3 people inside
- in study room number 1, next to the library with 15 children inside
- in study room number 2, next to the library, room that is connected to the hall in front of the school library
- in the hall of the school
- in the medical ward, with 8 people inside the room.

In the school in Giurgiu, investigations were made

- in a classroom at 8:00 a.m.
- in a classroom at 11:00 a.m.
- in a classroom at 1:00 p.m.
- in the school hall.

6.1.3. Offices

Fungi were investigated in eight offices.

6.1.4. Archives

Fungi were investigated in 10 archive rooms, where the possibility of contamination was presumed.

6.1.5. Shops

The levels of bacteria in air were investigated (a) in a supermarket in Bucharest City with ground floor and 4 floors and (b) in a shop with only a ground floor.

The investigations were made in the morning when the shop opens and in the evening when the shops were nearly to be closed.

6.2. Sampling methods

For the bacteriological and fungal evaluation, the sampling methods used were a passive method and an active one [26, 27].

6.2.1. Passive sampling method

The total number of mesophilic aerobic bacteria, yeast, and molds in the air of selected rooms was determined using the Koch sedimentation method. Bacteria from air were harvested using the gravitational method on the Petri plates with specific growth media. The measurements of levels of bacteria in air were made in several points on each location. The results were expressed as CFU/m^3. The equation used for calculating the results was as follows:

$$CFU/m^3 = n \times 10{,}000/s \cdot t/5$$

where n is the number of colonies on the Petri plate, s is the surface of the Petri plate, and *t* is the time of Petri plate exposure. Petri dishes with growth media were exposed for 10 min.

Results obtained by the Koch sedimentation method are less accurate than those from impaction methods with the use of an air sampler. The sedimentation method does not require expensive instrumentation and is cheap and simple. The sedimentation method does not permit exact quantitative determination. Still, this method permit an assessment of the types of microorganisms present in the air and can give an approximation of bacterial and fungal concentration in the air of investigated indoor spaces.

6.2.2. Active sampling method

This method used an impactor sampler M.A.S.Q.-90, with 2 l/min airflow and a total air volume equal to 200 l.

The following growth media were used [28]:

- tryptic soy agar (TSA) to determine the total number of bacteria
- blood agar to determine the presence of *Staphylococcus* and *Streptococcus*
- Czapek Dox Agar (Oxoid) for fungi identification

To obtain the total number of bacteria in the air of investigated rooms, the Petri dishes with growth media were incubated at 37°C for 24 h and then left for another 24 h at room temperature. To obtain the number of fungi colonies, the Petri dishes were exposed for 10 days at 25°C.

The identification of bacteria types was made by macroscopic estimation (description of colony) and by microscopic estimation.

Diagnosis of filamentous fungi was based on estimation of morphological features of growth on Czapek medium as well as on microscopic observation.

6.3. Indicators used to assessed the air contamination of indoor air

The following were assessed and counted on a Petri dish:

a. The total number of bacteria, expressed in CFU/m^3 (colony-forming units/cubic meter, CFU/m^3)

b. The presence of *Streptococcus* alpha and beta hemolytic

c. The presence of *Staphylococcus aureus* and non-*aureus* with and without hemolytic proprieties

d. The presence of coliform bacteria as an indicator of enteric contamination of air

e. The total number of fungi expressed in CFU/m^3 as quantitative assessment of contamination

f. The types of fungi as a qualitative assessment of indoor air contamination with fungi

6.4. References values

To have a good assessment of the results that were obtained, it was useful to have as references standardized values for the biological contamination of indoor air.

At present, there are no standards or guidelines for microbiological quality of indoor air. Each country has individual regulations and recommendations.

The Commission of the European Communities (CEC) defines the following level of bacteriological air contamination:0 undetectable, 1-499 CFU/m^3 low, 500-999 CFU/m^3 medium, >1, 000 CFU/m^3 high, and 2, 000 CFU/m^3 very high. A value of 10, 000 CFU/m^3 of nondescript airborne microbes could therefore be considered a hazardous level for indoor environment [12].

In 2001, the American Industrial Hygiene Association (AIHA) published a proposition of guidelines that stipulate 1, 000 of viable CFU/m^3 in air as an upper limit for concentration in indoor environment and for the amount of fungal spores in different indoor environments. It is proposed that the indoor air levels of microorganisms should be less than 500 CFU/m^3 for residential buildings and less than 250 CFU/m^3 for commercial buildings [29, 30].

According to current Swedish requirements, the number of 500 colony-forming units (CFU) of bacteria and 300 CFU of fungal spores in 1 m^3 can be accepted in an indoor environment [31].

The regulation in Brazil stipulates that the upper limit for total amount of airborne microorganisms (especially fungi) should be 750 CFU/m^3 [32, 33].

In Hong Kong, the indoor air levels of microorganism should be less than 1, 000 CFU/m^3 of bacteria. The levels are considered excellent if they are less than 500 CFU/m^3 [34, 35].

In Taiwan, the indoor air levels of bacteria should not exceeded 500 CFU/m^3 and 1, 000 CFU/m^3 for fungi [36, 37].

In India, there are no guidelines for the levels of bacteriological contamination of air [38].

In Singapore, the regulation for indoor air quality imposes an upper limit for the concentration of bacteria of 500 CFU/m^3 [39].

The American Conference of Governmental Industrial Hygienists (ACGIH) stipulates that the counts for total bacteria should not exceed 500 CFU/m^3[40].

WHO (1988) has set a limitation of fungi counts at 500 CFU/m^3 [41].

The Korean Indoor Air Quality Control in Public Use Facilities Act prescribes a total airborne bacterial concentration below 800 CFU/m^3 for hospitals, nurseries, public nurseries, and hospitals for the elderly and post partum care centers [42, 43].

In Romania, the reference value for indoor air contamination in a house is equal to 2, 500 CFU/m^3 and 550 CFU/m^3 for fungi. In schools, the level of bacteriological contamination of air should not exceed 1, 500 CFU/m^3 for bacteria at 37°. The presence of coliform bacteria and hemolytic bacteria are not allowed in air inside education units.

7. Results

7.1. Houses in Bucharest City

The highest value for air contamination with germs at 37°C was equal to 21, 466 CFU/m^3 in one of the apartments with one room (apartment B).

In the kitchens, the highest value of contamination identified was equal to 12, 701 CFU/m^3, the minimum value was equal to 201 CFU/m^3, and the highest average value was equal 3, 200 CFU/m^3 (Fig. 1).

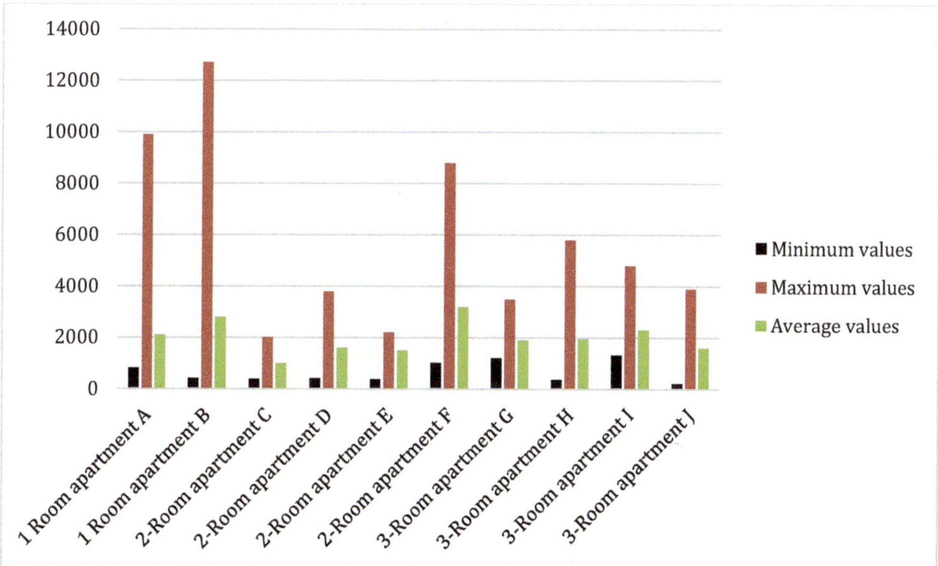

Figure 1. Bacteriological contamination of kitchen air (CFU/m^3 at 37°C).

In the living rooms, the highest identified value of contamination was equal to 21, 466 CFU/m³, the minimum value was equal to 210 CFU/m³, and the highest average value was equal 4, 731 CFU/m³ (Fig. 2).

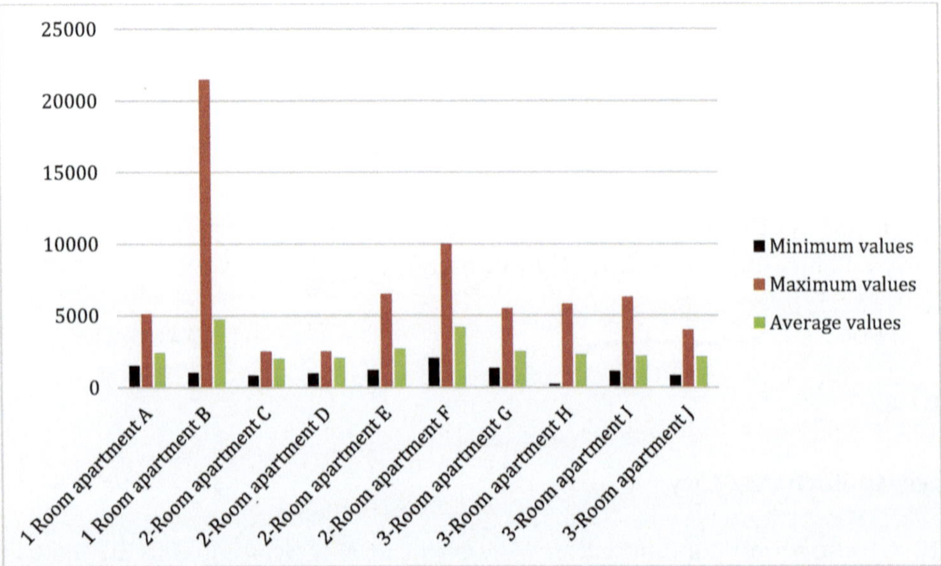

Figure 2. Levels of bacteriological air contamination in living rooms (CFU/m³ at 37°C).

Taking into account the moment of the day, the highest bacteriological contamination (average values) was found in the living room, in the evening, and was equal to 6, 665 CFU/m³ in the one room apartment—apartment B (Fig. 3).

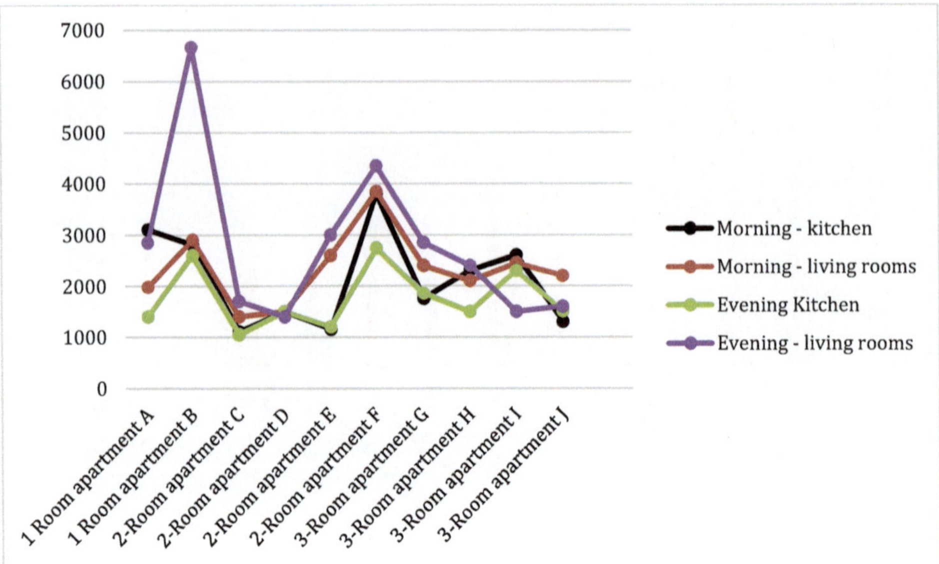

Figure 3. Average numbers of germs at 37°C in the morning and in the evening in investigated apartments.

The indoor air contamination is higher in the fall and in the winter, reaching a level equal to 7, 853 CFU/m³ in one room apartment (average values)—apartment B (Fig. 4).

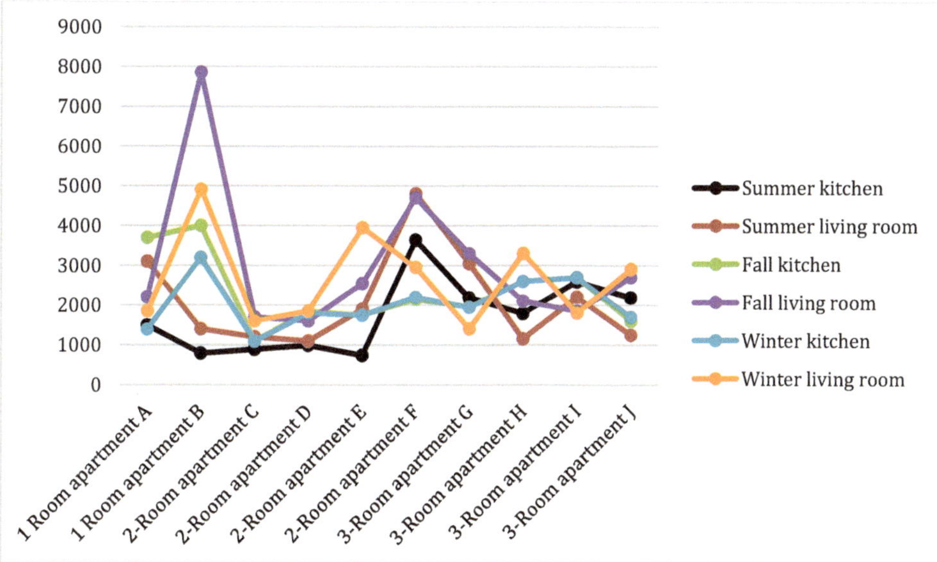

Figure 4. Average numbers of germs at 37°C according to the season in investigated apartments.

In connection to the smoking habit of the inhabitants, apartments with one and two rooms had a higher bacteriological contamination of air (Fig. 5).

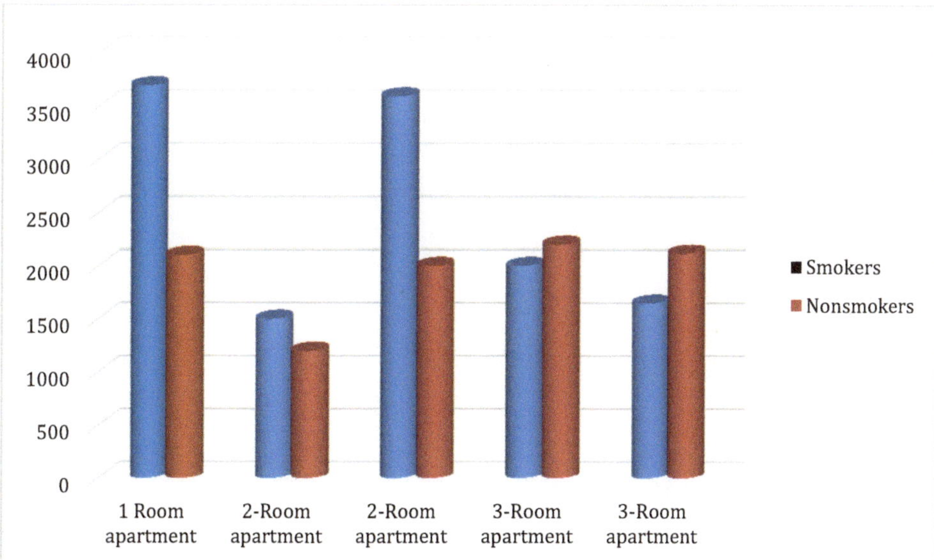

Figure 5. Number of germs at 37°C in houses according to the smoking habit.

In connection to the type of apartment, bacteriological contamination is higher in the apartments with two linked rooms. Bacteriological contamination is also presented in the three-room apartment (Figs. 6 and 7).

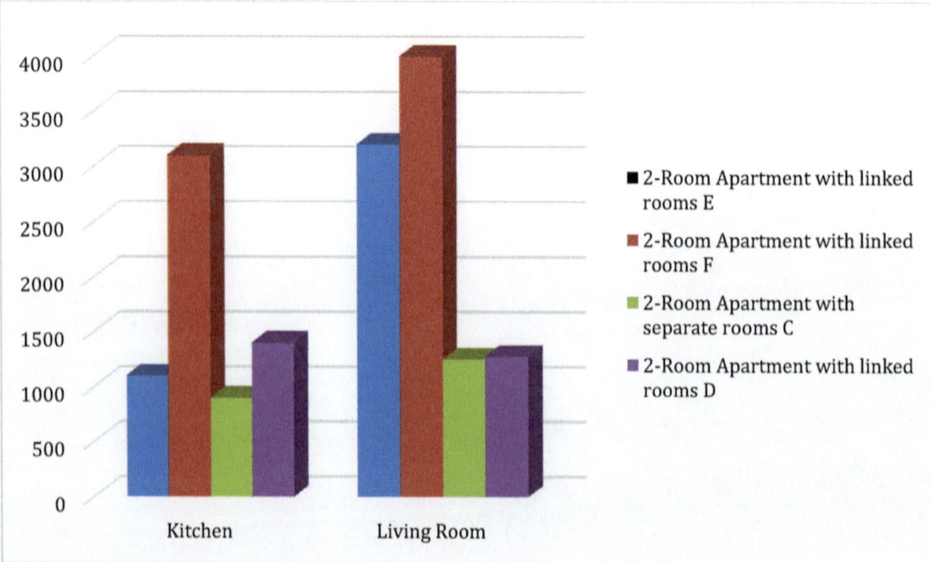

Figure 6. Number of germs at 37°C and the type of apartment (apartments with two rooms).

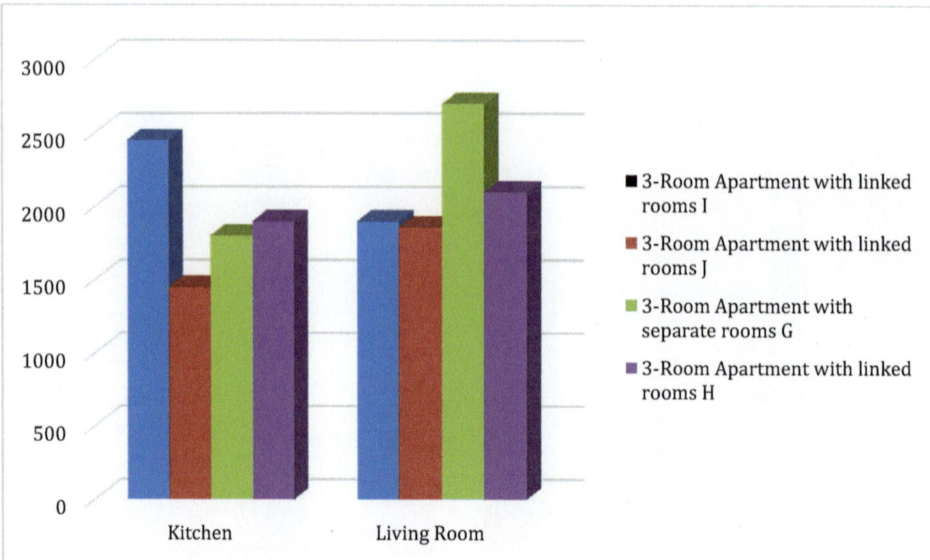

Figure 7. Number of germs at 37°C and the type of apartment (apartments with three rooms).

Contamination of air with staphylococci reached the highest value equal to 7, 988 CFU/m^3 in the kitchen and 10, 157 CFU/m^3 in the living room (Figs. 8 and 9).

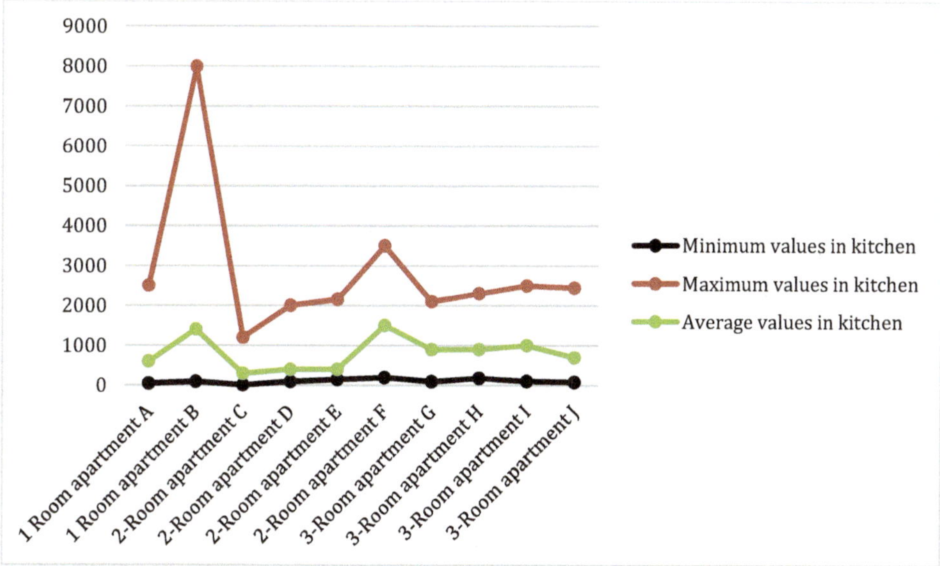

Figure 8. Contamination with staphylococci in kitchen air (CFU/m³).

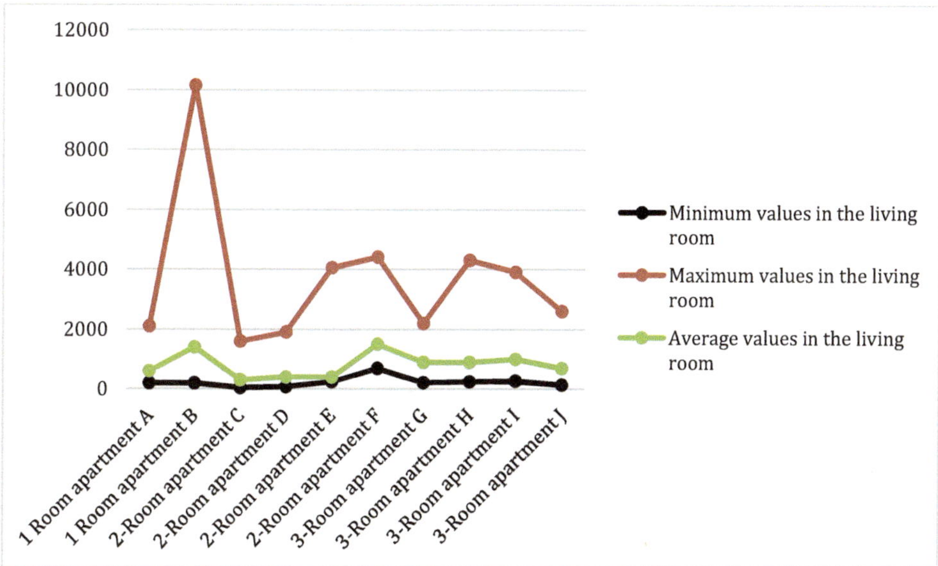

Figure 9. Contamination with staphylococci in living rooms air (CFU/m).

The ratio between staphylococci/germs at 37°C is around 40% in the kitchen and over 50% in the living rooms (Fig. 10).

The assessment of the fungal contamination showed that in many homes, the level of fungal contamination was very high in all investigated rooms. The levels of fungi were between 635 and 10, 859 CFU/m³. The highest values measured were 2, 802 CFU/m³ in the living room, 2,

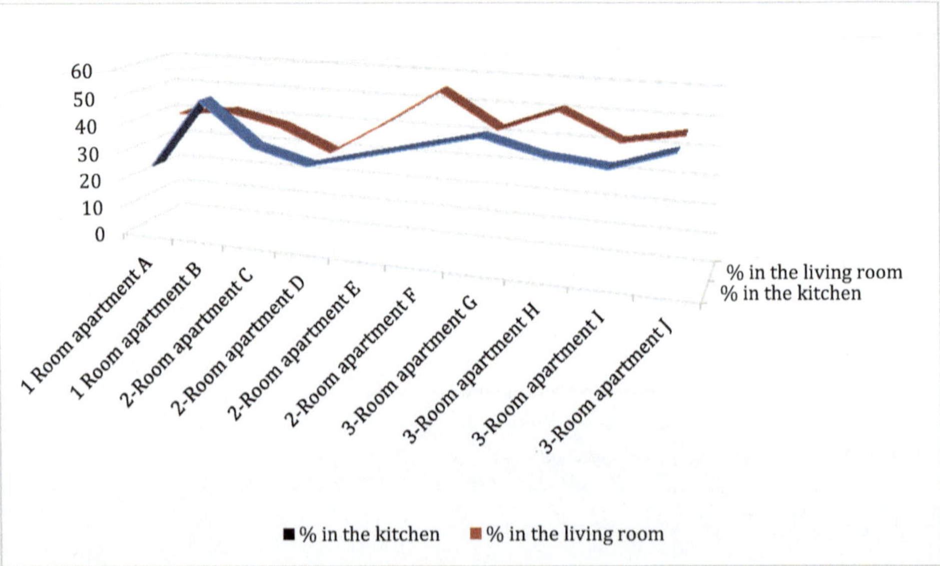

Figure 10. Ratio between staphylococci/germs at 37° in investigated houses.

302 CFU/m^3 in the kitchen, 7, 675 CFU/m^3 in the bathroom, and 10, 859 CFU/m^3 in the bedroom (highest overall). The species of fungi identified were *Penicillium*, *Aspergillus niger*, *Rhizopus*, *Mucor*, *Cladosporium herbarum*, and *Fusarium*.

7.2. Schools

In a school in Bucharest, the following results were obtained: in an empty classroom, before 8:00 a.m., the value identified was 25 CFU/m^3; during the school day (at 9.30:00 a.m.) in a classroom with 23 children, the value identified was 67 CFU/m^3; in a classroom with 36 children and 1 teacher, the value identified was 229 CFU/m^3; in a classroom with 23 children and 1 teacher, the value identified was 140 CFU/m^3; in a classroom (at 11:00 a.m.) with 24 children and 1 teacher, the value identified was 1, 465 CFU/m^3 with 38 CFU/m^3 *Streptococcus* alpha-hemolytic present; in the school library (at 10:00 a.m.) with 3 people inside, the biological contamination of air was equal to 45 CFU/m^3; in study room number 1, next to the library with 15 children inside, the level of air contamination was equal to 89 CFU/m^3; in study room number 2, next to the library, the room connected to the hall in front of the school library, the level of biological air contamination was equal to 427 CFU/m^3; the level of air contamination in the school hall was equal to 180 CFU/m^3; in the medical ward, with 8 people inside the room, the level of biological air contamination was equal to 60 CFU/m^3, 1 CFU/m^3 *Streptococcus* alpha hemolytic, and 1 CFU/m^3 *Streptococcus* beta hemolytic present (Fig.11).

In a school in Giurgiu City, the following results were obtained: at 8:00 a.m., a number of 1, 259 CFU/m^3 was identified; at 11:00 a.m., a number of 1, 732 CFU/m^3 was identified; at 1:00 p.m., a level of biological air contamination equal to 472 CFU/m^3 was identified. In the school hall, the level of air contamination was equal to 2, 202 CFU/m^3, the level of contamination of the air with S. non-*aureus* was equal to 1, 259 CFU/m^3 at 8:00 a.m. and 11:00 a.m., and 472 CFU/m^3 at 1:00 p.m.; the level of contamination with *Streptococcus* alpha-hemolytic was equal to 157

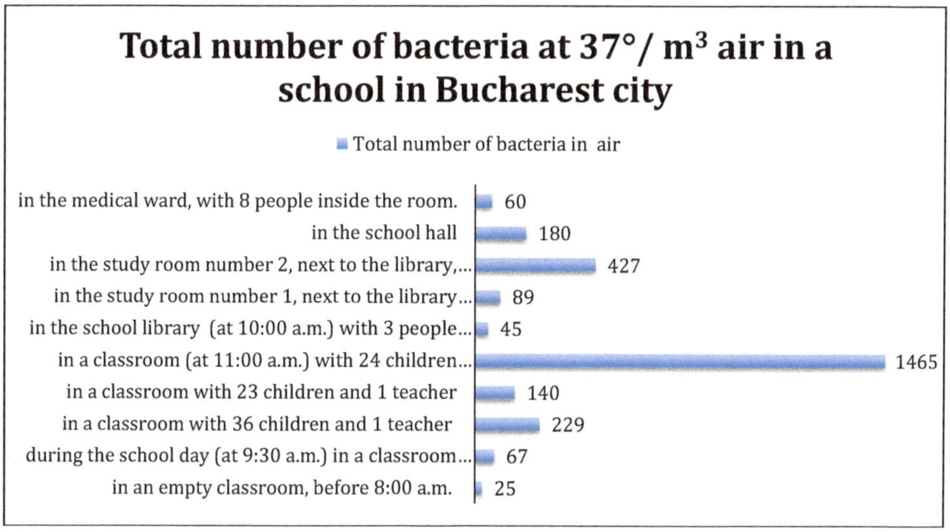

Figure 11. Level of bacteriological contamination of air in a school in Bucharest City.

CFU/m³ at 11:00 a.m. In a classroom and in the school hall, at 1:00 p.m., coliform bacteria were identified as an indicator of enteric contamination of air. The level of contamination with coliform bacteria was equal to 315 CFU/m³. The detected levels of coliform bacteria can be dangerous for children health (Fig.12).

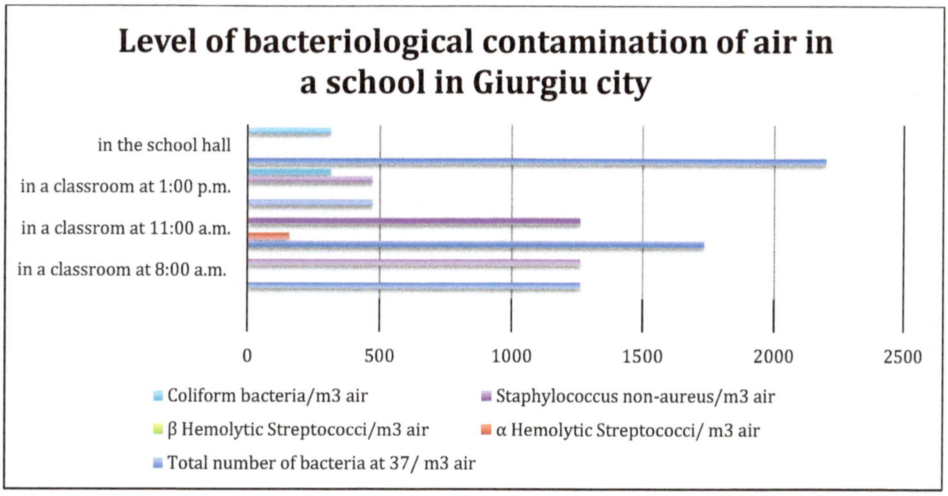

Figure 12. Levels of bacteriological air contamination in a school in Giurgiu City.

7.3. Offices

The levels of fungal contamination were between 734 and 4, 034 CFU/m³. The fungi species identified were *Penicillium* spp., *A. niger*, *A. flavus*, *Aspergillus versicolor*, *Mucor*, and *Cladosporium*.

7.4. Archives

The levels of fungal contamination were between 655 and 1, 490 CFU/m^3. By a qualitative assessment of air pollution with fungi, the following types of fungi were identified: *Penicillium* spp., *A. niger, A. flavus, A. versicolor, Mucor, Cladosporium.*

7.5. Shops

In a supermarket with ground floor and 4 floors, the levels of bacteriological contamination are shown in Table 1.

Average values		Total number of bacteria at 37°/m^3 air	Total number of staphylococci/m^3 air	Hemolytic staphylococci/m^3 air	*Staphylococcus aureus*/m^3 *air*	α Hemolytic streptococci/m^3 air	β Hemolytic streptococci/m^3 air
Ground floor	M	5, 140.75	3, 296	161.25	197.72	328.75	3.24
	E	13, 836.00	10, 397.25	404.5	240.75	964.25	0
First floor	M	3, 919.25	2, 015.5	137.5	170.55	213.66	0
	E	9, 258.75	7, 316	372	232.5	569.33	0
Second floor	M	6, 155.45	4, 276.33	299.75	226.25	182.5	0
	E	10, 916.75	8, 705.5	339	286.25	549.16	0
Third floor	M	6, 946	4, 914.25	280.25	141	273.5	0
	E	13, 509.25	10, 800.5	419.5	316	775.5	0
Fourth floor	M	6, 473.75	4, 292.75	301.25	189.97	228	0
	E	10, 352.25	7, 759.75	405.25	267.78	560	0

Table 1. Levels of bacteriological air contamination in a supermarket in Bucharest City (M = morning, E = evening)

The highest levels of contamination were registered in the evening and were equal to 13, 836 bacteria at 37°/m^3 air at the ground floor and 13, 509 bacteria at 37°/m^3 air at the third floor. In the second and fourth floors, in the evening, the level of bacteria in air were equal to 10, 917 bacteria at 37°/m^3 air and 10, 352 bacteria at 37°/m^3 air. In the first floor, in the evening, levels of bacterial contamination equal to 9, 259 bacteria at 37°/m^3 air were registered. The levels of air contamination were the lowest in the first floor (Fig. 13).

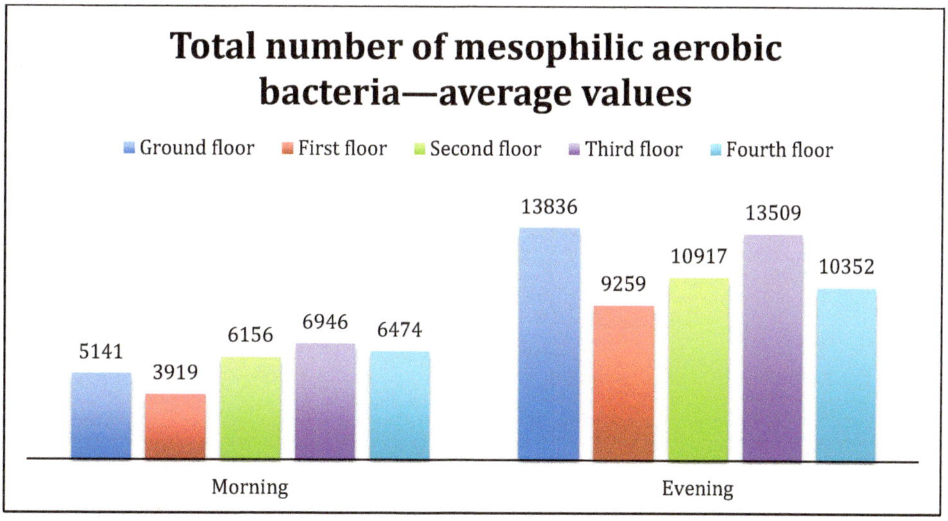

Figure 13. Levels of bacteriological contamination of air in a supermarket—average values.

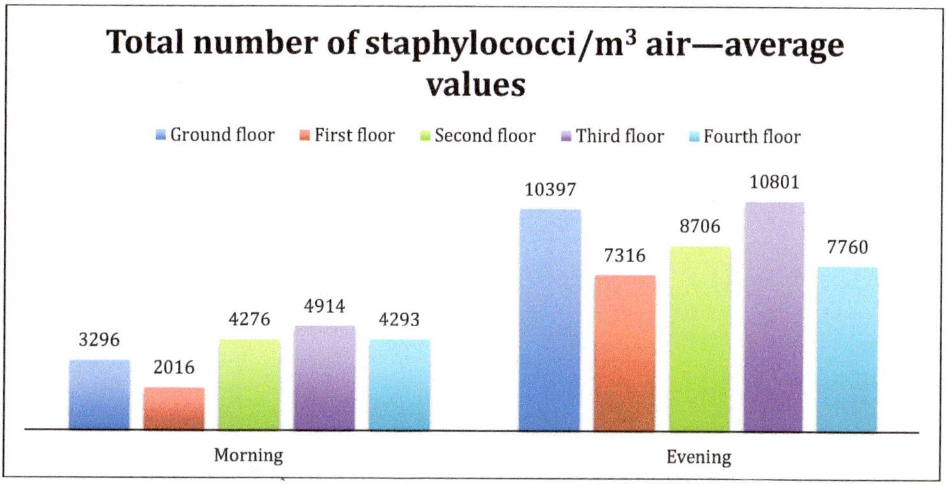

Figure 14. Total number of staphylococci/m^3 air—average values.

The contamination of air, in the case of the supermarket with a ground floor and other four floors, with staphylococci was high in the evening. The highest levels were registered in the third floor with a contamination burden equal to 10, 801/m^3 air, followed by the ground floor with 10, 397/m^3 air, the second floor with 8, 706/m^3 air, the fourth floor with 7, 760/m^3 air, and the first floor with 7, 316/m^3 air. It can be seen that the levels of air contamination were the lowest in the first floor (Fig.14).

Hemolytic staphylococci/m^3 air and *S. aureus*/m^3 air were identified in the same amount in the evening at al floors.

α Hemolytic streptococci/m^3 air were identified in the highest levels in the evening at the ground floor. In the evening, at the other floors, the levels were almost equal.

β Hemolytic streptococci/m^3 air were identified at the ground floor only.

In a shop with only a ground floor, the levels of bacteriological contamination are presented in Table 2.

Average values		Total number of bacteria at 37°/m^3 air	Total number of staphylococci/m^3 air	Hemolytic staphylococci/m^3 air	*Staphylococcus aureus/m^3 air*	α Hemolytic streptococci/m^3 air	β Hemolytic streptococci/m^3 air
Ground floor	M	2, 398.16	1, 323.2	78	47.3	86.5	0
	E	3, 107.5	2, 217.85	90.08	164.76	75.65	0

Table 2. Levels of bacteriological air contamination in a shop with only a ground floor in Bucharest City (M = morning, E = evening)

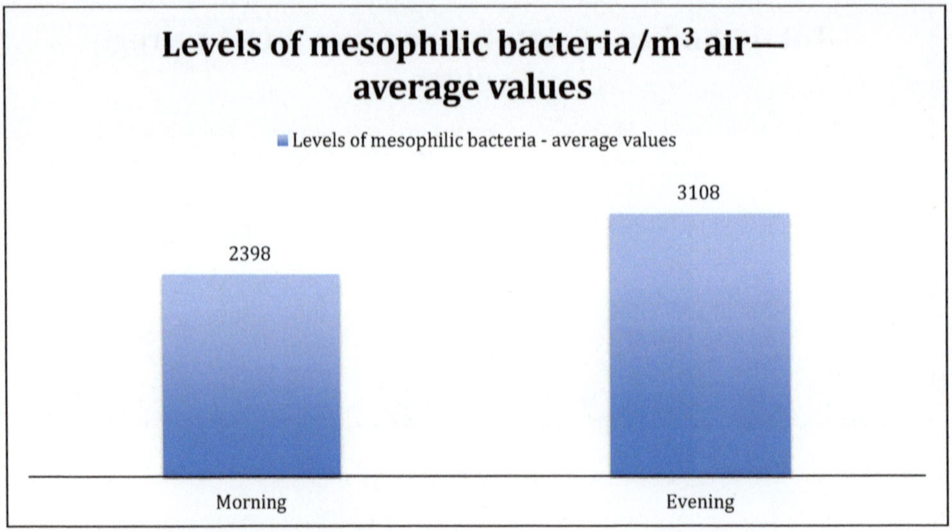

Figure 15. Levels of bacteriological contamination of air in shop with only a ground floor.

In the shop with only a ground floor, the levels of air contamination registered higher values in the evening than in the morning measurements for mesophilic bacteria, total number of staphylococci, hemolytic staphylococci, *S. aureus*, and α hemolytic streptococci (Fig.15).

8. Discussions

In the investigated houses, the highest value for air contamination with germs, at 37°C, was found in one of the apartments with one room (apartment B). Regarding the time of the day, the highest level of bacteriological contamination was found in the evening, in the one room apartment—apartment B. Taking into account the time of the year, the indoor air contaminations were higher in the fall and winter, especially in the one room apartment—apartment B.

Considering the smoking habit of the inhabitants, apartments with one and two rooms had a higher bacteriological contamination of air where there were smokers present.

Considering the type of the apartments, in apartments with more than one room, bacteriological contamination is at the highest level in the apartment with two linked rooms.

The ratio between staphylococci/germs at 37°C is around 40% in the kitchen and over 50% in the living rooms.

In a previous study [44], the levels of bacteriological and fungal pollution of indoor air were as follows:

In summer, the maximum levels of bacteria were equal to 57, 500 CFU/m^3 in the kitchen and bathroom and 32, 500 CFU/m^3 in the living room with average values between 6, 090 and 13, 241 CFU/m^3.

In winter, the maximum levels of bacteria were equal to 12, 600 CFU/m^3 in the kitchen, 50, 400 CFU/m^3 in the bathroom, and 6, 300 CFU/m^3 in the living room with average values between 2, 406 and 7, 903 CFU/m^3.

The highest levels of staphylococci in air were, in the summer, between 26, 125 CFU/m^3 in the living room and 55, 000 CFU/m^3 in the bathroom with average values between 4, 308 CFU/m^3 and 7, 200 CFU/m^3. In winter, the maximum values were between 10, 250 CFU/m^3 in the kitchen and 55, 000 CFU/m^3 in the bathroom, with the average values between 3, 219 and 6, 421 CFU/m^3.

The air pollution with fungi was visualized as a layer on the Petri dish in the kitchen and bathroom, and in the living room the maximum level registered was equal to 1, 820 CFU/m^3. The fungi species identified were the same like in the present study.

Another study performed in Torun University in Poland [45] had the following results:

- The levels of air contamination with bacteria were between 7 and 3, 313 CFU/m^3 as the background and between 53 and 5, 673 CFU/m^3 air at different sample site. The amount of bacteria found in the outdoor air ranged from 10 to 310 CFU/m^3.
- The air pollution with fungi ranged between 0 and 893 CFU/m^3 in the morning and 0 and 1, 373 CFU/m^3 in the afternoon. The outdoor fungal background ranged between 217 and 3, 750 CFU/m^3.

Another study that investigated the levels of biological contamination of indoor air in eight libraries of the Jimma University in Addis Ababa [46] had the following results:

- The levels of bacteria were between 367 and 2, 595 CFU/m^3 with an average value equal to 1, 476. CFU/m^3.
- The levels of fungi were between 524 and 1, 992 CFU/m^3 with an average value equal to 1, 087 CFU/m^3.

- The isolated bacteria were Micrococcus sp., S. aureus, Streptococcus pyogenes, Bacillus sp., and Neisseria sp.
- The isolated fungi were *Cladosporium* sp., *Alternaria* sp., *Penicillium* sp., and *Aspergillus* sp.

The microbiological quality of air in the investigated rooms in the Bucharest school as well as in the Giurgiu school changed significantly during the school day. The highest levels were found around 11:00 a.m. in classrooms with large numbers of persons inside. Most of the levels identified in the study were under the limit of 1, 000 CFU/m^3, with the exceptions of one measurement in the school in Bucharest and four measurements in the school in Giurgiu City. Also, the presence of *Streptococcus* beta hemolytic and coliform bacteria in classrooms air represents a threat to the health of children. This threat appears to be greater in the school in Giurgiu City.

The fungi levels identified in offices and archives were over the limit accepted in Romania. Isolated fungi species include *Cladosporium, A. flavus, A. niger,* and *Penicillium* sp.

In the case of the shop with only a ground floor and the supermarket, it is obvious that the size of the shop, the numbers of floors, and the numbers of departments in the shop can induce a high bacteriological contamination of air due to the high number of people visiting the shops. The air contamination in the supermarket was found to be higher at the ground floor and at the third and fourth floors. The high level of bacteriological contamination at the ground floor is probably due to the large number of visitors, while at the third and fourth floor it is probably due to the accumulation of bacteria coming from the first and second floors. In the shop with only a ground floor, the levels of air contamination in the evening were lower than in the morning. It is probable that this type of shop has better ventilation. High levels of staphylococci and the presence of *Streptococcus* in air can be dangerous for human health, and it is necessary to have a very strong disinfection action.

9. Conclusions

From this study, several conclusions can be drawn:

a. Biological contamination (bacteriological and fungal contamination) represents a constant element that was identified in all Romanian houses.

b. All results showed that the levels of bacteria in indoor air are higher than the allowed limit (2, 500 CFU/m^3 air for bacteria at 37°—limits admitted in Romania for rooms in a house). In schools, the levels of bacteriological contamination of air must not exceed 1, 500 CFU/m^3 for bacteria at 37°. The presence of coliform bacteria and hemolytic bacteria are not allowed in the air inside education units).

c. Concerning the indoor air pollution with fungi, the results showed a high level of contamination exceeding the allowed limit in Romania of 550 CFU/m^3 for fungi in air.

d. Biological contamination of air in schools can represent a threat for children health. This situation is obvious especially in schools outside Bucharest.

e. In work environment like offices or archives, the presence of fungi may represent a health risk for those who are exposed during the work process.

f. In the shops, a high biological contamination can be identified not only due to the numerous visitors but also by the numerous vendors present inside. The employees can be exposed during the working days to an important bacteriological load.

g. The hygienic quality of indoor air is revealed by the levels of biological air contaminants (bacteria and fungi).

h. A poor biological quality of air in all investigated environments was identified in the present study.

i. A constant exposure to biological contaminants (bacteria and fungi) from air can induce specific pathologies like asthma, allergic diseases, digestive diseases, and other infectious diseases. Among the isolated fungi, there were strongly allergenic and toxic species. The increased level of fungal flora in insufficiently ventilated rooms could be a reason for serious health problems of people occupying and working in those rooms.

j. All the investigated places require measurement to ensure good ventilation and proper measures of hygiene.

Author details

Anca Maria Moldoveanu[1,2]

Address all correspondence to: anca.moldoveanu@gmail.com

1 Carol Davila University of Medicine and Pharmacy, Bucharest, Romania

2 National Institute of Public Health, Bucharest, Romania

References

[1] Nevalainen A, Morawska Lidia, Biological Agents in Indoor Environments Assessment of Health Risks, Work conducted by a WHO Expert Group between 2000-2003. Edited by Aino Nevalainen and Lidia Morawska (2009).

[2] WHO Guidelines for Indoor Air Quality: Dampness and Mould, World Health Organization, Copenhagen, Denmark (2009).

[3] World Health Organization. Indoor Air Pollutants: Exposure and Health Effects, Copenhagen, WHO Regional office for Europe (EURO Reports and Studies No 78) (1983).

[4] Park JH, Spiegelman DL, Gold DR, Burge HA, Milton DK. Predictors of airborne endotoxin in the home. Environ Health Persp (2001), 109:859-64.

[5] Bischof W, Koch A, Gehring U, Fahlbusch B, Wichmann HE, Heinrich J. Predictors of high endotoxin concentrations in the settled dust of German homes. Indoor Air (2002), 12(1):2-9.

[6] Baun-Fahrlander C, Riedler J, Herz U, Eder W, Waser M, Grize L, Maisch S, Carr D, Gerlach F, Bufe A, Lauener RP, Schierl R, Renz H, Nowak D, Von Mutius Erika. Environmental exposure to endotoxin and its relation to asthma in school-age children. N Engl J Med (2002), 347(12):869-77.

[7] Waser M, Schierl R, Erika Von Mutius, Maisch S, Carr D, Riedler J, Eder W. Schreuer M, Nowak D, Braun-Fahrländer C. Determinants of endotoxin levels in living environments of farmers' children and their peers from rural areas. Clin Exp Allergy (2004), 34:389-397.

[8] Park J-H, Spiegelman DL, Burge HA, Gold DR, Chew GL, Milton DK. Longitudinal study of dust and airborne, endotoxin. Environ Health Perspect (2000), 108(11): 1023-1028.

[9] Douwes J. (1-3)-β-D-glucans and respiratory health: a review of the scientific evidence. Indoor Air (2005), 15: 160-169.

[10] Douwes J, Zuidhof A, Doekes G, van der Zee S, Wouters I, Boezen HM, Brunekreef B. (1 → 3)-β-D-glucans and endotoxin in house dust and peak flow variability in children. Am J Respir Crit Care Med (2000), 162: 1348-1354.

[11] Haalen Khan AA, Karuppayil SM. Fungal pollution of indoor environments and its management. Saudi J Biol Sci (Oct 2012), 19(4): 405-426.

[12] European Collaborative Center—Indoor Air Quality & Its Impact On Man, Report No. 12, Biological Particles in Indoor Environment, Luxembourg (1993).

[13] Institute of Medicine (IOM). National Academy of Sciences, Committee on the Assessment of Asthma and Indoor Air. Clearing the Air: Asthma and Indoor Air Exposures. National Academy Press, Washington DC (2000).

[14] Simpson JC, Niven RM, Pickering CA, Fletcher AM, Oldham LA, Francis HM. Prevalence and predictors of work related respiratory symptoms in workers exposed to organic dusts. Occup Environ Med (1998), 55(10):668-672.

[15] Flamant-Hulin M, Annesi-Maesano I, Caillaud D. Relationships between molds and asthma suggesting non-allergic mechanisms. A rural-urban comparison, Pediatr Allergy Immunol (2013), 24: 345-351.

[16] Watcharoot K, Mendell MJ, Gaffin JM, Wang G, Phipatanakul W. Indoor environmental exposures and exacerbation of asthma: an update to the 2000 review by the Institute of Medicine. Environ Health Perspect (January 2015), 123(1): 6-20.

[17] Mendell MJ, Mirer Anna G, Kerry Cheung Kerry, Tong My, Douwes J. Respiratory and allergic health effects of dampness, mold, and dampness-related agents: a review of the epidemiologic evidence. Environ Health Perspect (2011), 119:748-756.

[18] Annesi-Maesano I. Indoor exposition to molds and health outcome. Rev Allergy Clin Immunol (2013), 23; 21-25.

[19] Simpson JC, Niven RM, Pickering CA, Fletcher AM, Oldham LA, Francis HM. Prevalence and predictors of work related respiratory symptoms in workers exposed to organic dusts. Occup Environ Med (1998), 55(10):668-672.

[20] Fung F, Hughson WG. Health effects of indoor fungal bioaerosol exposure. Proc Indoor Air 2002, 46-51.

[21] Joshi SM. The sick building syndrome. Indian J Occup Environ Med (Aug 2008), 12(2): 61-64.

[22] Shoemaker RC, House DE. Sick building syndrome (SBS) and exposure to water-damaged buildings: time series study, clinical trial and mechanisms, Neurotoxicol Teratol (2006), 28(5):573-588, Epub 2006.

[23] Fraczek K, Górny RL. Microbial air quality at Szczawnica sanatorium, Poland. Ann Agric Environ Med (2011), 18, 63-71.

[24] Madureira J, Paciência I, Rufo JC, Pereira C, Teixeira JP, de Oliveira FE. Assessment and determinants of airborne bacterial and fungal concentrations in different indoor environments: homes, child day-care centres, primary schools and elderly care centres. Atmos. Environ (2015), 109: 139e146.

[25] Solomon J. Leviston-Porter School District—Annual Microbiological Assessment. Leviston-Porter Elementary School, New York (2005).

[26] Stryjakowska-Sekulska M, Piotraszewska-Pajak A, Szyszka A, Nowicki M, Filipiak M. Microbiological quality of indoor air in university rooms. Polish J Environ Stud (2007), 16(4): 623-632.

[27] Méheust D, Le Cann P, Reboux G, Millon L, Gangneux J-P. Indoor fungal contamination: health risks and measurement methods in hospitals, homes and workplaces, http://informahealthcare.com/mby. ISSN: 1040-841X (print), 1549-7828 (electronic). Crit Rev Microbiol. Online: 2013 Informa Healthcare USA.

[28] Ulla L. Characterization of microorganisms in indoor environments. Publications of the National Public Health Institute—Academic Dissertation, Kuopio (2008).

[29] IRSST, Occupational Health and Safety Research Institute Robert Sauvé: Bioaerosol in Workplace: Evaluation, Control and Prevention Guide. Technical guide, 2001, www.irsst.qc.ca.

[30] AIHA. Field Guide for the Determination of Biological Contaminates in Environmental Samples. Fairfax, VA: American Industrial Hygiene Association (1996).

[31] Abel E, Andersson JV, Dawidowicz N, Christophersen E, Hanssen SO, Lindén A-L, Lindvall T, Pasanen AL. The Swedish key action 'The Healthy Building'—research results achieved during the first three years period 1998-2000, Indoor Air 2002. Proceedings of the 9th International Conference on Indoor Air Quality and Climate (2002), 2:996-1001.

[32] de Aquino Neto FR, de GóesSiqueira LF. Guidelines for indoor air quality in offices in Brazil. Proceedings of Healthy Buildings (2000), 4:549-554.

[33] Nunes ZG, Martins AS, Altoe ALF, Nishikawa MM, Leite MO, Aguiar PF, Fracalanzza SEL. Indoor air microbiological evaluation of offices, hospitals, industries, and shopping centers. Mem Inst Oswaldo Cruz (2005), Vol. 100(4): 351-357.

[34] Lee SC, Li W-M, Ao C-H. Investigation of indoor air quality at residential homes in Hong Kong—case study. Atmos Environ (2002), 36:225-237.

[35] Li WM, Lee SC, Chan LY. Indoor air quality at nine shopping malls in Hong Kong. Sci Total Environ. (Jun 2001), 12;273(1-3):27-40.

[36] Huang P-Y, Shi Z-Y, Chen C-H, Den W, Huang H-M, Tsai J-J. Airborne and surface-bound microbial contamination in two intensive care units of a medical center in Central Taiwan. Aerosol Air Qual Res (2013), 13: 1060-1069.

[37] Hsu C-H, Lu M-C, Huang D-J. Application of chlorine dioxide for disinfection of student health centers. Environ Monit Assess (2012) 184:741-747.

[38] Katiyar V. Assessment of indoor air micro-flora in selected schools. Adv Environ Res (2013), 2(1): 61-80.

[39] Obbard JP, Fang LS. Airborne concentrations of bacteria in a hospital environment in Singapore. Water Air Soil Pollut (2003), 144 (1): 333-341.

[40] American Conference of Governmental Industrial Hygienists (ACGIH). Guidelines for the Assessment of Bioaerosols in the Indoor Environment. ISBN-13: 978-0936712833 (1989).

[41] World Health Organization (WHO). Indoor Air quality: Biological Contaminants: European Series, Number No. 31. WHO Regional Publication, Copenhagen (1988).

[42] Ministry of Environment (MOE). Korean Indoor Air Quality Control in Public Use Facilities, etc. Act. Republic of Korea (2011).

[43] Lee JH, Ahn KH., Yu IJ. Outbreak of bioaerosols with continuous use of humidifier in apartment room. Toxicol Res (June 2012), 28(2): 103-106.

[44] Moldoveanu AM. Indoor air pollution in Romanian homes. In: Environmental Technologies—New Developments. ISBN: 978-3-902613-10-3. I-Tech Education and Publishing, Wien (2008).

[45] Kalwasińska A, Burkowska A, Wilk I. Microbial air contamination in indoor environment of the university library. Ann Agric Environ Med (2012); 19(1): 25-29.

[46] Hayleeyesus SF, Manaye AM. Microbiological quality of indoor air in university libraries. Asian Pac J Trop Biomed (May 2014), 4(Suppl 1): S312-S317.

6

Air Pollution in Welding Processes — Assessment and Control Methods

Farideh Golbabaei and Monireh Khadem

Additional information is available at the end of the chapter

1. Introduction

Welding is a very common operation in many industries and workplaces [1, 2]. According to American Welding Society, it is defined as "a metal joining process wherein coalescence is produced by heating to suitable temperature with or without the use of filler metal" [3]. There is a variety of welding processes that are used in different working conditions. According to some reports, from 0.2 to 2.0% of the working population in industrialized countries are engaged in welding activities [4]. Worldwide, over five million workers perform welding as a full time or part time duty [5, 6]. These welders, depending on conditions, work in outdoor or indoor workplaces, in open or confined spaces, underwater, and above construction sites. Welding operators face various hazards resulting in different injuries, adverse health effects, discomfort and even death. Furthermore, air pollution due to welding leads to certain consequents on humans and environment. Therefore, there are strong reasons to deal with the welding processes and the working environment of the welder from different aspects. A large number of welders experience some type of adverse health effects. Other workers near the place where welding process is done may be affected by the risks generated by it [1, 7]. Totally, welding risks can be classified as risks deriving from physical agents and risks related to the chemical components. The main hazards related to welding include electricity, radiation, heat, flames, fire, explosion, noise, welding fumes, fuel gases, inert gases, gas mixtures and solvents. Welders may be exposed to other hazards not directly related to welding, such as manual handling, working at height, in confined spaces, or in wet, hot or humid situations, and working with moving equipment, machinery and vehicles. Welding in a static awkward or horizontal posture may result in musculoskeletal injuries, such as strains and sprains. Prolonged use of a hard hat and a helmet can cause strain on the neck. Furthermore, long-term exposure, repetitive motions with arms and hands, and tasks inducing high force may lead to cumulative effects, increasing risk of injury. The main components of welding emissions are oxides of metals due to contact between the oxygen in the air and the vaporized metals. Common chemical hazards include particulates (lead, nickel, zinc, iron oxide, copper, cadmium, fluorides, manganese, and chromium) and gases (carbon monoxide, oxides of nitrogen, and ozone). Recently, nanoparticles emitted by welding operations are considered as an important group of air pollutants and there is a need to assess particle sizes and size distributions when risk assessment is done. Each welding technique produces a distinctive range of particulate composition and morphology. Different and complex profiles of exposures may be related to various welding environments [8-10].

HAZARD	WELDING PROCESS			
	PAW/PAC Carbon Arc Processes	SMAW GTAW GMAW FCAW	SAW	Oxyfuel
Ergonomic	+	+	+	+
Electric Shock	+	+	+	x
Bright light	+	+	-	+
Ultraviolet radiation	+	+	-	x
Toxic fumes and gases	+	+	-	+
Heat, Fire, and Burns	+	+	+	+
Noise	+	x	x	x

x No hazards, + Hazard present, - Hazard present if SAW flux is absent [11]

Table 1. The hazards associated with welding Processes

2. Welding technology

2.1. Applications

Welding is used extensively in various manufacturing industries including shipyards, automobile factories, machines, home appliances, computer components, bridge building and other constructions. Welding is used for manufacturing pressure vessels, heat exchangers, tanks, sheet metal, prefabricated metal buildings and architectural work. Also, welding is an applicable technique in maintenance operations and repair shops. It is used in mining, oil and gas transmission companies, piping systems, heavy equipment manufacturing, aerospace, electronics, medical products, precision instruments, electric power, and petrochemical industries. Perhaps artists and sculptors are the smallest group who use welding techniques to create artworks. Therefore, many things that people use in daily lives are welded or made by welded parts [12].

2.2. Workplace conditions

Welders, depending on conditions, work in outdoor or indoor workplaces, in open or confined spaces, underwater, and above construction sites. In some conditions, welding processes are carried out in confined spaces where the welding work area is surrounded on most sides by walls and there is no sufficient space for the installation of a conventional exhaust hood [1, 7].

Working in indoor environments includes all works which are done in buildings like workshops, repairing shops, storages, office, and any closed area in industries, factories, and other places. Welders may work in indoor areas to do welding tasks full time or part time. An important benefit of indoor workplaces is the protection against environmental factors such as rain, wind and sunshine. Outdoor workers spend long periods of time working in open areas. They are exposed to different hazards depending on their type of work, as well as geographic region, season, and the period of time they are outside. Outdoor works include agriculture, construction, mining, oil and gas transmission through pipelines, transportation, warehousing, utilities, and service sectors. Sometimes welders should work in such workplaces to do their tasks. Some workplace hazards related to outdoor areas include unpredictable weather conditions, bugs and wild animals, extreme heat, extreme cold, and ultraviolet (UV) radiation.

Many workplaces contain spaces that are considered "confined" because their configurations hinder the activities of employees who must enter, work in, and exit them. A confined space has limited or restricted means for entry or exit. Confined spaces include underground vaults, tanks, storage bins, manholes, reactor vessels, silos, process vessels, and pipelines. Confined spaces have the following characteristics: limited space, entry, or exit; poor ventilation and lack of safe breathing air. Welders may experience various hazards when welding in confined spaces, such as fire, explosion, electric shock, asphyxiation, and exposure to hazardous air contaminants [13-16].

2.3. Types of welding processes

There are different welding processes (over 50 types) that differ greatly in some parameters such as heat, pressure, and the type of equipment used. Welding process can be classified into various types based on different literatures. Some common types of welding are listed in five categories each of which includes some subcategories (Figure 1). The most common and known types of welding include:

Shielded Metal Arc Welding: (SMAW) also is known as Manual Metal Arc welding (MMA) or stick electrode welding. It is one of the oldest, simplest, and most versatile arc welding processes used for carbon steel welding and low alloy welding. In SMAW, the electrode is held manually, and the electric arc flows between the electrode and the base metal. The electrode is covered with a flux material which provides a shielding gas for the weld to help minimize impurities. A wide range of metals, welding positions and electrodes are available based on intended requirements. This type of welding is especially suitable for jobs such as the erection of structures, construction, shipbuilding, and pipeline work. Contrary to the other methods requiring shielding gas which are unsuitable in wind, SMAW can be used outdoors in different weather conditions. However, owing to the time required for removing the slag after welding and changing the electrodes, its arc time factor is relatively low. As a disadvantage, forming fumes in SMAW makes the process control difficult.

Gas Metal Arc Welding: (GMAW) or metal inert gas (MIG) welding is used for most types of metal and is faster than SMAW. It may be applied to weld vehicles, pressure vessels, cranes, bridges and others. This process involves the flow of an electric arc between the base metal and a continuous and consumable wire electrode. Shielding gas (usually an argon and carbon dioxide mixture) is supplied externally; therefore, the electrode has no flux coating or core. MIG welding is used for mild steel, low alloyed and stainless steel, for aluminum, for copper, nickel, and their alloys. Some parameters can affect MIG welding process, such as:

- Electrode diameter
- Voltage
- Wire feed speed and current
- Welding speed
- Shielding gas and gas flow rate
- Torch and joint position

To perform an optimum welding, most of the mentioned parameters should be matched to each other. In addition to affecting the quality of welding, some of these parameters can influence the fumes and gases emitted from the process. However, the fume produced by MIG welding is less than that of SMAW. Unlike the SMAW that is discontinuous due to limited length of the electrodes, GMAW is a continuous welding process. There is no slag and no need for high level of operators' skill. Nevertheless, expensive and non-portable equipment is required, and also outdoor applications are limited because of the negative effects of weather conditions like wind on the shielding gas [17, 18].

Gas Tungsten Arc Welding: (GTAW) is also known as tungsten inert gas (TIG) welding. GTAW is used on metals such as aluminum, magnesium, carbon steel, stainless steel, brass, silver and copper-nickel alloys. This technique uses a permanent non-consumable tungsten electrode. The filler metal is fed manually, the weld pool and the electrode are protected by an inert gas (usually argon), and high electrical currents are used in this type. Welding of stainless steel, welding of light metals, such as aluminum and magnesium alloys, and the welding of copper are the main applications of TIG welding. GTAW welds are highly resistant to corrosion and cracking over long time periods. However, TIG welding is suitable to weld thin materials and produces a high quality weld of most of metals. There is no need for slag removal in GTAW process. The concentration of heat takes place in a small zone, resulting in the minimal thermal distortion of work piece. The TIG welding has some disadvantages including low welding rate, expensiveness, and need for high level of operators skill. Although during TIG welding operators are exposed to dangerous gases and fumes, the generation of these compounds is very little in comparison with other welding processes.

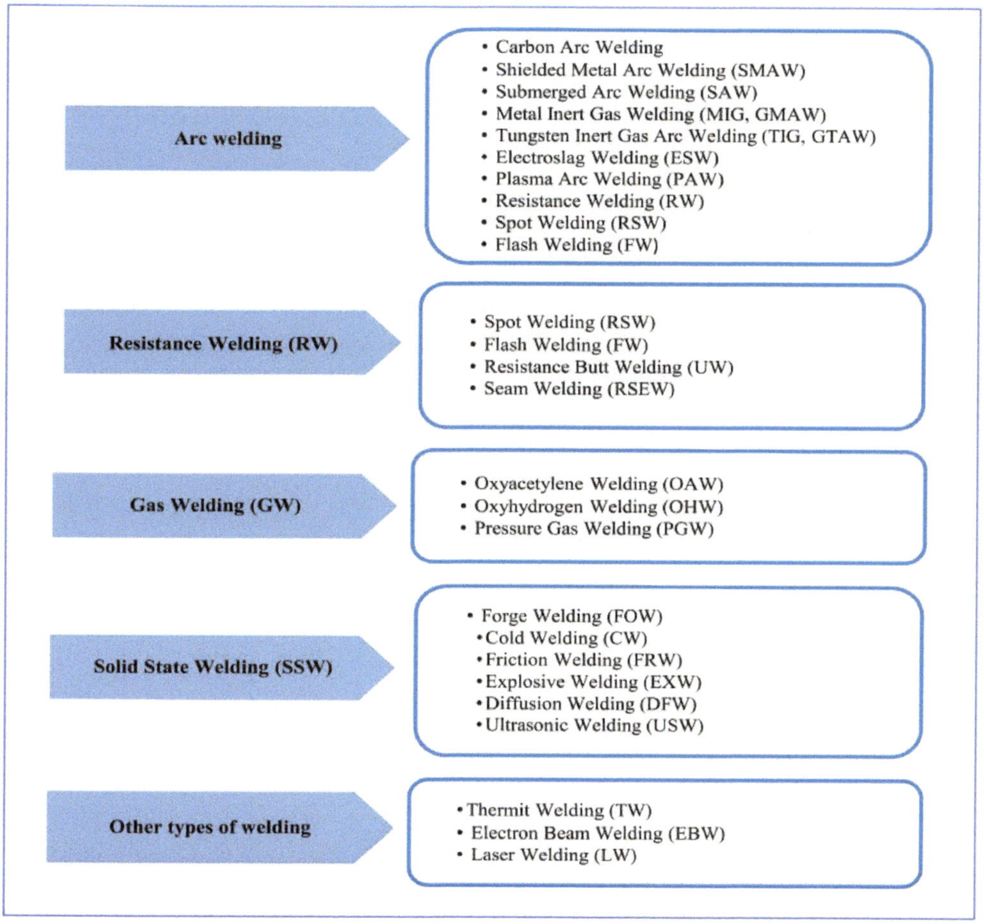

Figure 1. Classification of welding processes [18]

Submerged Arc Welding: (SAW) is a highly-productive welding method (4-10 times as much as the SMAW). SAW may be automatic or semi-automatic. It is used to weld thick plates of carbon steel and low alloy steels. In this welding process, the electric arc flows between the base metal and a consumable wire electrode; however, the arc is not visible since it is submerged under flux material. This welding process is usually used for large structures such as large tubes, cylindrical vessels, and plates in shipyards. Some parameters can affect SAW process such as welding arc voltage, arc current, the size and shape of the welding wire, and the number of welding wires. A low fume emission is produced during SAW process and there is a little ozone, nitric oxide and nitrogen dioxide generation because of the invisibility of the arc. Very high welding rate, suitability for automation, suitability for both indoor and outdoor works, and high weld quality are mentioned as advantages of SAW. Some limitations of this welding process include: slag inclusion, limited applications often for welding in a horizontal position, and need for precise parameter setting and positioning of the wire electrode.

Plasma Arc Welding: (PAW) is an arc welding process in which arc is formed between an electrode and the workpiece. In PAW process, the plasma arc can be separated from the shielding gas cover by positioning the electrode within the body of the torch. It can be named as a key difference between GTAW and PAW. Two inert gases are used in the process, one forms the arc plasma and the second shields the arc plasma. Applying the plasma arc welding is being increased in industries, because it provides a high level of control and accuracy to produce high quality welds. Also, using the PAW leads to long electrode life for high production conditions. This welding process is suitable for both manual and automatic applications. It can be used for precise welding of surgical equipment, jet engine blades, and instruments required for food and dairy industry. There is a low level of fume generation during PAW, but welding gases especially ozone is often formed in this process. Need for less operator skill, high welding rate, high penetrating capacity, long electrode life, high accuracy and precision, and short weld time are considered as the advantages of PAW process. Its limitations include expensive process tools, needs for high power electrical equipment, more distortion and loss of mechanical properties due to the greater heat input.

Flux Core Arc Welding: (FCAW) is used for carbon steels, low alloy steels and stainless steels. This welding process has similarities to both SMAW and GMAW. This process is used in construction because of its high welding speed and portability. The consumable tubular electrode is continuously fed from a spool and an electric arc flows between the electrode and base metal. The electrode wire has a central core containing fluxing agents. There are a variety of cored wires; some of them require the use of shielding gas like carbon dioxide or the mixture of argon/carbon dioxide and the others (self-shielded flux cored wires) do not require additional shielding gas. The slag produced in FCAW process acts as an additional protection during cooling time but has to be chipped away after that. Like other welding process, FCAW has some advantages and limitations. No needs for skilled operators and pre-cleaning of metals, suitability for use in the outdoor or windy condition (it is true about self-shielded flux cored wires), suitability for use in all positions, and ease of varying the alloying constituents are mentioned as FCAW advantages. Its limitations include: emission of considerable amount of fumes in self-shielded wires, higher price of filler material and wire in comparison with GMAW, and needs for slag removal. Also, escaping of the shielding gas from the welded area leaves holes in welded metal, resulting in porosity in products [17, 18].

3. Air pollution out of welding

According to Flagan and Seinfeld definition, "the phenomenon of air pollution involves a sequence of events: the generation of pollutants at and their release from a source; their transport and transformation in and removal from the atmosphere; and their effects on human beings, materials, and ecosystems" [19]. Air pollution is indoor or outdoor contamination by particulates, biological molecules, or other harmful materials that changes the natural characteristics of the Earth's atmosphere. Household combustion devices, motor vehicles, forest fires, and industrial processes are common sources of air pollution. Major industrial sources of particulate matter include the metals, mineral products, petroleum, and chemical industries. Air pollution is considered as a threat to human health as well as to the Earth's ecosystems. Based on WHO report, around 7 million people worldwide died due to the air pollution in

2012 [20]. Welding, as an important operation in most industries, can considerably cause air pollution. In all types of welding processes, fume and gases are formed as air pollutants. Due to high temperature during the welding process, different substances in the arc are vaporized. Then, the vapor condenses and oxidizes in contact with the air, leading to the formation of fumes. The fume particles are so small and they can reach the narrowest airways of respiratory system (respiratory bronchioles). Some parameters like the welding type and consumables (filler metal and surface coatings) determine the kind and amount of generated particles and gases.

The composition of welding fumes and their generation rate is a function of different parameters. Welding fume particles are in the fine (<2.5 μm) to ultrafine (<100 nm) respirable size and can penetrate into the alveolar regions of the lungs. The generation of fumes depends on:

- -Amperage, voltage, gas and arc temperatures and heat input in the welding process
- -Consumables like electrodes
- -Materials
- -Welding duration [9, 21].

The most common gases emitted during welding are ozone, nitrous gases and carbon monoxide. Phosphine and phosgene are the other gases that may be produced during welding. Gases are generated due to the high temperature and ultraviolet (UV) radiation from the arc. Like fumes, some factors can affect the emission of gases during welding processes. For instance, ozone formation during welding depends on process type, used material, and shielding gases. Welding gases can also be generated when surface coatings or contaminants contact with hot surfaces or UV radiation.

Along with harming human health, air pollution may lead to various environmental impacts. Air pollution can adversely cause critical impacts on the atmosphere and natural environment in many ways. Welding, as an industrial process, causes serious impacts on the environment depending on its operation mode and the technological equipment. Environmental pollution in welding process is the result of some parameters, such as high percentage of heat that is released into the environment and materials including large amount of gases and fumes. Some factors needed to carry out the welding operation include: energy, mineral or organic substances (protective gases, cooling water, oils, grease and protective substances etc.). These consumables can be harmful for the environment. Furthermore, produced waste during the welding processes results in undesirable impact on the work or natural environment. To protect the welding region and prevent oxidation, inert gases like carbon dioxide and argon are used because of their availability and low cost. They are used as shielding gases and have undesirable impacts on the environment. To protect the environment and keep the resource for future, energy conservation and reducing greenhouse gas emissions should be considered. In this respect, the average consumption rate, usage rate and the purity of products and consumables are important factors [22, 23].

The generation of fumes and gases is directly related to the welding process. Fumes emitted during manual metal arc welding (MMA) and MIG welding is the same. In some conditions, the level of fume generated during MIG welding (with solid wire) may be much lower in comparison with the fumes produced by MMA. In TIG welding, a lower level of fumes is emitted compared to MMA and MIG welding. The composition of fumes is directly associated with the composition of used wire. MMA welding causes adverse health effects because of forming the hexavalent chromium (Cr (VI)) in the process. In addition, high rates of emission of toxic compounds generate in MMA-stainless steel (MMA-SS) welding [24]. During TIG welding, very little fume are generated. Welding fumes may be composed of oxides of chromium, nickel and copper, with very low specific limit values. The individual elements and also their synergetic effect must be considered when assessing fume toxicity. Lower ozone and nitrogen oxides are emitted during TIG welding than those in MIG/MAG welding. The amount of mentioned gases during TIG welding is dependent on current, arc length and the flow and type of shielding gas. High electrical currents cause the significant levels of ozone, nitric oxide and nitrogen dioxide. During MIG welding, significant levels of ozone and nitrogen oxides are produced because of intense current levels.

There is a little information concerning emissions during plasma arc welding (PAW). Due to the similarity of TIG and PAW welding techniques, they may probably emit air pollutants with the same magnitude. MIG welding of aluminum produces larger quantities of ozone than TIG welding of aluminum. Forming more nitrogen oxides in the latter process will keep the emitted ozone levels down [25, 26]. A study by Schoonover et al. showed that welders performing MIG and SMAW are exposed to higher fume concentrations than welders performing TIG. According to mentioned study, exposure to manganese during MIG was nearly two and ten times higher than in SMAW and TIG, respectively. In fact, not using a consumable electrode during TIG welding results in lower exposures. The highest average exposures occur in SMAW, followed by GMAW, and GTAW [21]. K. Fuglsang et al. investigated the Fume Generation Rates (FGR). This rate for MMA was 3-5 times higher than that found for MAG and MIG. The same FGR was found for TIG and MIG/MAG welding [27].

Various welding processes generate particles in different size distributions. Particles produced during MMAW, MAG, MIG, and laser welding are quite similar in size. Resistance Spot Welding (RSW) and TIG welding have a completely different structure for particle size distribution. These techniques produce particles smaller than 100 nm, in which, at least 90% are smaller than 50 nm. Particles generated during processes with high mass emission rates (MMAW, MAG, MIG, and Laser) have diameters about 100–200 nm and there are few nanoscaled particles between them. Processes with low mass emission rates (TIG and RSW) generate exclusively particles smaller than 50 nm; however, the number concentration of particles in these techniques is similar to the others. Although, welding types with low mass emission rates are called "clean techniques", their potential toxicological properties and health effects due to exposure to nanoscaled particles should be further studied [28].

A study by Keane M. introduced the pulsed axial spray method (from MIG process) as the best choice of the welding processes because of minimal fume generation (especially Cr (VI)) and cost per weld. The advantages of this method include usability in any position, high metal deposition rate, and simple learning and use. Totally, the highest amounts of fume are produced by the self-shielded cored wire electrodes. These electrodes are used without a shielding gas. Using solid wire electrodes results in emission of ozone and nitrogen oxides as in MAG welding [25, 29].

Airborne particles with diameter smaller than 100 nm are known as nanoparticles or ultrafine particles. According to researches, nanoparticles are more harmful to human health than larger particles. They can deeply penetrate inside the respiratory system and then enter the blood stream. The main character of nanoparticles is the high surface area, and their toxicity depends on the shape and penetration potential inside the respiratory system. In addition to the emission of fine particles with diameter less than 10 μm, nanoparticles may be emitted during welding operations. Some studies have indicated that the highest values of nanoparticles are related to MAG and TIG processes when applying the highest current intensities. Therefore, the higher amounts of nanoparticles are emitted by processes in which the higher energy intensities are used.

As it was stated, the emission of nanoparticles during welding operations increases with the increase of welding parameters like current intensity. Welding with short-circuit mode results in lower value of nanoparticles, because its low current intensity and tension causes an electric arc with lower temperature and thus emitting lower amounts of elements. Also, the high quantity of nanoparticles is generated by the stainless steel welding, which can be related to the presence of helium in the gas mixture of welding. Helium, due to high ionization energy, results in electric arc with high temperature that generates higher values of nanoparticles. Furthermore, the study of different base materials indicated that the higher quantity of nanosized particles is obtained for stainless steel compared to carbon steel. According to data from different investigations, the lowest level of ultrafine particles deposited in alveolar region of lungs was related to FSW, followed by TIG and MAG. Totally, all welding processes can result in deposition of a significant concentration of nanosized particles in lungs of exposed welders [30-32].

4. Welding health effects

Fume and gases emitted during welding pose a threat to human health while welding. The exposures may be varied depending on where the welding is done (on the ship, in confined space, workshop, or in the open air). The welding process and metal welded affect the contents of welding fumes. On the other hand, physical and chemical properties of the fumes and individual worker factors are effective on deposition of inhaled particles. In this respect, particle size and density, shape and penetrability, surface area, electrostatic charge, and hygroscopicity are the important physical properties. Also, the acidity or alkalinity of the inhaled particles are the chemical properties that may influence the response of respiratory tract. Welding gases can be classified into two groups; some gases are used as a shielding gas and the others are generated by the process. Shielding gases are usually inert, therefore, they are not defined as hazardous to health but they may be asphyxiants. Gases generated by welding processes are different based on welding type and may cause various health effects if over-exposure occurs. Welding emissions depending on some factors like their concentration, their properties, and exposure duration can lead to health effects on different parts of human body.

Hazards on Respiratory System - The inhalation exposures may lead to acute or chronic respiratory diseases in all welding processes. In the occupational lung diseases, the various reactions produced in respiratory tract depend on some parameters such as the nature of the inhaled matter, size, shape and concentration of particles, duration of exposure, and the individual workers susceptibility. Chronic bronchitis, interstitial lung disease, asthma, pneumoconiosis, lung cancer, and lung functions abnormalities are some hazardous effects on respiratory systems. The pulmonary disorders are various based on the differences in welding metals and their concentrations. Ozone, at low concentrations, irritates the pulmonary system and can cause shortness of breath, wheezing, and tightness in the chest. More severe exposures to ozone can lead to pulmonary edema. Exposure to nitrogen dioxide may cause lung function disorders like decrements in the peak expiratory [33, 34]. Kim JY in a study showed the PM2.5 concentration for welders (1.66 mg/m^3) was significantly greater than that for controls (0.04 mg/m^3), and the exposure of healthy working population to high levels of welding fumes resulted in the acute systemic inflammation [35].

Hazards on Kidney- Substantial exposure to metals and solvents may be nephrocarcinogenic. Chromium can deteriorate renal function because of accumulation in the epithelial cells of the proximal renal tubules and induce tubular necrosis and interstitial changes in animals and humans. Tubular dysfunctions have been identified in subjects occupationally exposed to Cr (VI) [33, 36]. Welders exposed to heavy metals like cadmium and nickel have also experienced kidney damage [7]. Pesch et al. indicated that there was an excess nephrocarcinogenic risk involved with soldering, welding, milling in females. So, it can be considered an evidence for a gender-specific susceptibility of the kidneys [37].

Hazards on Skin - Erythema, pterygium, non-melanocytic skin cancer, and malignant melanoma are the adverse health effects of welding on the skin among which erythema is a common one. The intense UV as well as visible and infrared radiations are produced by welding arc machines. Exposure to UV can lead to short- and long-term injuries to the skin [33, 38-40]. Some metals like beryllium, chromium and cobalt can cause direct effects (irritation and allergic impacts) on the skin. Also, they may be absorbed through the skin and cause other health effects such as lung damage. When the particles are small and there are cuts or other damages to the skin, the absorption through the skin is raised [7, 36]. Chromium (VI) may cause irritating and ulcerating effects when contacting with skin. An allergic response including eczema and dermatitis may be induced in sensitized individuals exposed to Cr (VI) [34].

Hazards on the visual systems - Most welding processes emit intense ultraviolet as well as visible and infrared radiations. Adverse effects on the eyes may be induced by these optical radiations. In addition, Tenkak reported that, welding may cause photokeratitis and some types of cataract. Erhabor et al. showed the most frequent symptoms among the welders were eye irritation (95.43%). Exposure to UV radiation can lead to short- and long-term injures to the eyes. Acute overexposure to UV radiation can result in the photokeratitis and photoconjunctivitis that are the inflammation of the cornea and the conjunctiva, respectively. These responses of the human eye to UV radiation are commonly known as snow blindness or welder's flash [33, 38, 41].

Hazards on Reproductive System - In the past, some studies have indicated the increased risk for infertility and reduced fertility rate in mild steel welders. There are some evidences that reduced fecundity can be related to exposure to hexavalent chromium and nickel. According to new investigations, damages to male reproduction system have been reported less than before, probably because of decreasing the exposure levels in the developed countries. However, some special tasks like stainless steel welding may impair welders' reproduction system [42-44]. A study by Bonde showed that mild steel welding, but not stainless one, resulted in significant effects on the fertility during years [45]. Mortensen [46] observed a greater risk for poor sperm quality among welders compared to controls, especially welders who worked with stainless steel. Therefore, welding in general, and specifically with stainless steel, may cause the reduced sperm quality. According to Sheiner, impaired semen parameters can be associated with the exposures to lead and mercury [47].

Hazards on the nervous system - Memory loss, jerking, ataxia and neurofibrillary degeneration have been attributed to exposure to aluminum. The accumulation of aluminum in the brain may develop some neuropathological conditions, including amyotrophic lateral sclerosis, Parkinsonian dementia, dialysis encephalopathy and senile plaques of Alzheimer's disease [36]. A review of literatures by Iregren suggests that occupational exposure to manganese results in the central nervous system damage that is generally irreversible [48]. Although there are multiple toxic agents in welding, more literatures have dealt with manganese as an important agent of toxicity. Welders are also exposed to high concentrations of carbon monoxide and nitrogen dioxide. Carbon monoxide can cause the neurological impairments of memory, attention, and visual evoked potentials. Both central and peripheral nervous system damages may be induced by exposure to welding fumes [49]. Some neurobehavioral impairments associated with exposure to lead and manganese have been indicated by Wang [50]. A study by Bowler (2003) showed there is a relation between welding and a decline in brain functions and motor abilities. In this survey, various questionnaire and tests like neuropsychological tests were used [49].

Carcinogenic effects - There are some concerns regarding the presence of carcinogens in the welding fumes and gases. Sufficient evidences for carcinogenicity of nickel, cadmium, and chromium (VI) have been reported through experimental and epidemiological studies. These three metals have been categorized as carcinogen "Class 1" by the International Agency for Research on Cancer [51-52]. Ozone has been introduced as a suspect lung carcinogen in experimental animals, but there are very few documents about its long term effects on welders. The ultraviolet emissions resulting from welding arc can potentially cause skin tumors in animals and in overexposed individuals, however, there is no definitive evidence for this effect in welders [53].

Other health problems - Welding on surfaces covered with asbestos insulation may lead to risk of asbestosis, lung cancer, mesothelioma, and other asbestos-related diseases in exposed welders. The intense heat and sparks of welding can cause burns. Eye injuries are possible because of contact with hot slag, metal chips, and hot electrodes. Lifting or moving heavy objects, awkward postures, and repetitive motions result in strains, sprains and musculoskeletal disorders. High prevalence of musculoskeletal complaints (back injuries, shoulder pain, tendonitis, carpal tunnel syndrome, and white finger) is seen in welders [54].

5. Exposure standards for welding emissions

Usually, exposure standards apply to long term exposure to a substance over an eight hour work per day for a normal working week, over an entire working life. Some organizations like American Conference of Governmental Industrial Hygienists (ACGIH), National Institute for Occupational Safety and Health (NIOSH), and Occupational Safety and Health Administration (OSHA) have published the exposure standards for various components in welding fumes and gases (table 2). According to Work Safe Australia exposure standards cannot be used as a fine dividing line between a healthy and unhealthy workplace. Adverse health effects below the exposure limits might be seen in some people because of individual susceptibilities and natural biological variation. ACGIH, however, recommends a TLV-TWA (Threshold Limit Value-Time Weighted Average) of 5 mg/m^3 for total welding fume, assuming that it contains no highly toxic components. Each metal or gas within the welding has its own exposure standard. As Table 2 indicates, biological media, Biological Exposure Index (BEI), and carcinogenicity class have been proposed for some welding emissions [55, 56].

Substance	OSHA PEL-TWA (mg/m3)	NIOSH REL-TWA (mg/m3)	ACGIH TLV-TWA (mg/m3)	ACGIH BEI	Carcinogenicity
Aluminum Fume	15 (Total) 5 (res)	5	5		
Arsenic	0.01	0.002 (Ceiling)	0.01	35 μg As/L	A1
Barium	0.5	0.5	0.5		
Beryllium	0.002	0.5 (Ceiling)	0.002		A1
Cadmium Fume	0.005	LFC (Ca)	0.01 (Total) 0.002 (Res)	5 μg Cd/g creatinine	A2
Cobalt	0.1	0.05	0.02	15 μg Co/L	A3
Chromium(VI)	--	0.001	0.05	25 μg Cr/L	A1
Chromium metal	1	0.5	0.5		A4
Copper Fume	0.1	0.1	0.2		
Iron Oxide	10 (as Fe)	5	5		A4
Lithium	--	--	--		
Manganese	5 (Ceiling)	1	0.2	range 0.5 to 9.8 mg/L; up to 50 mg/L for occupational exposure	
Molybdenum	5(Soluble) 15 (Insoluble)	--	5 (Soluble) 10 (Insoluble)		

Substance	OSHA PEL-TWA (mg/m3)	NIOSH REL-TWA (mg/m3)	ACGIH TLV-TWA (mg/m3)	ACGIH BEI	Carcinogenicity
Lead	0.05	0.1	0.05	30 μg /dL (whole blood)	A3
Nickel	1	0.015 (Ca)	1	10μmol/mol creatinine	Elemental (A5) Insoluble inorganic (A1)
Platinum	0.002 (Soluble)	1(Metal) 0.002 (Soluble)	1		
Selenium	0.2	0.2	0.2		
Silver	0.01	0.01	0.1		
Tellurium	0.1	0.1	0.1		
Thallium	0.1	0.1(Soluble)	0.1	50 μg Th/g creatinine	
Titanium Dioxide	15	LFC (Ca)	10		
Vanadium Pentoxide	0.1 (Ceiling)	0.05(Ceiling)	0.05	50 μg V/g creatinine	
Zinc Oxide	5	5	5		
Zirconium	5	5	5		
Total fumes	--	LFC (Ca)	5		
Carbon monoxide	50 ppm	35 ppm	25 ppm	3.5% of (Hemoglobin) 20 ppm (end-exhaled air)	
Nitrogen dioxide	5 ppm (ceiling)	5 ppm (ceiling) 1ppm (STEL)	3 ppm		
Ozone	0.1 ppm	0.1 ppm	0.08 ppm		

LFC=lowest feasible concentration; Res=Respirable; Ca=NIOSH potential occupational carcinogen [55, 57, 58]

Table 2. Exposure limit of each individual constituent of welding components

6. Welding monitoring and risk assessment

6.1. Monitoring of welding emissions

Managing the risks of pollutants generated by welding process is carried out in some steps inculing identifying hazards, assessing the risks arising from these hazards, eliminating or minimising the risks via proper control ways, and checking the effectiveness of controls. Monitoring the welder's exposure is a main component of risk management process. Welding process leads to chemical exposures to fumes and toxic gases in enormous quantity. The hazard identification and risk assessment are necessary to work safely in a welding environment. Enough information, education, training and experience are required in this respect. In addition to the full-time welders, a large number of part-time welders who work in small shops and workers in the vicinity of the welding process may also be exposed.

There is a greater potential for exposure due to welding in confined spaces with poor ventilation such as ship hulls, metal tanks and pipe, therefore, monitoring such welders should be seriously considered.

As it was stated previously, the level of welder's exposure to welding emission depends on some factors like the process type, process parameters, and consumables used. Materials and consumables used in welding determine the chemical composition of welding emissions. The specific toxicity of each element and the synergetic effect of generated constituents must be considered to evaluate the exposure status of welders. There are some other workplace specific factors, including the ventilation condition, welder position or posture, and the volume of welding room, that influence the exposure level. The emission rate and also its concentration in the breathing zone of the welder or in the work environment are directly related to the mentioned factors. When it is probable that the welders' exposure will be exceeded the prescribed limits, or when the workers' health and the environment are at risk, the monitoring of hazards and the risk assessment program are required. To evaluate the hazards caused by different welding emissions, collecting various information is recommended. Air monitoring and measuring related pollutants via personal and environmental sampling, biological monitoring, workplace assessment with regard to physical and chemical hazards, and occupational medical findings can be used to evalute the welder's exposure status compeletely [59-60].

Air Monitoring -Airborne pollutants generated by welding can threaten the worker's health and safety. Thus, during the health and safety program, air monitoring is used to identify and quantify welding emissions. To evaluate air contaminants, a sampling strategy is used for collection of exposure measurements. The choice of the best strategy is based on site-specific conditions. In a sampling strategy, some parameters like selection of workers for personal monitoring, sampling duration and required number of samples are important. The measurement of contaminants is carried out in the breathing zone of selected worker. The collected samples must be representative of the normal work activity and exposure of welder, because the sampling results are used to prevent overexposures. Air monitoring in welding processes includes the sampling and analysis of welding fumes and welding gases [61].

Within recent years, standard practices have been developed to monitor exposures considering the occupational exposure limits for elements. Most measurements are made using personal monitoring systems with a pump at a proper flow rate connected to a cassette containing a membrane filter for a suitable period of time. To obtain the accurate result, filter cassette must be placed inside the welding helmet. Time-weighted average concentrations of total fumes is obtained by weighing the filter before and after exposure; the concentrations of elements are determined by chemical analysis methods provided by related organizations like American Welding Society and British Standards Institution [51], NIOSH Manual of Analytical Methods (NMAM) for metals in air and urine and OSHA Sampling and Analytical Methods are used to monitor the welding workplaces. In these methods, analysis of metals is performed by Inductively Coupled Argon Plasma-Atomic Emission Spectroscopy (ICP-AES) after sample preparation by acidic ashing [61, 62]. It is worth mentioning that the microwave digestion can be used instead of acidic ashing to prepare samples, leading to reduction in ashing time up to 90 percent, as well as cost saving and providing a healthier work environment for laboratory operators. Golbabaei et al. used the microwave digestion to prepare urine samples before urinary metal analysis by graphite furnace atomic absorption spectrometry [52].

As it was stated previously, there are different workplace conditions for workers who are welding in confined spaces compared to other welders. Limited access and little airflow or ventilation are the characteristics of a confined space. Hazardous concentrations of welding emissions can accumulate very quickly in such small spaces. Hazardous concentrations of welding emissions can accumulate very quickly in such small space. Thus, confined spaces should be monitored for toxic, flammable, or explosive emissions to evaluate welders' exposure. In some situations, continuous air monitoring may be necessary when workers are welding in a confined space with special conditions. Golbabaei et al. conducted an investigation to assess the risk related to welding pollutants for welders who work in confined spaces. Almost for all analyzed metals, there were significant differences between back welders and controls. Back welding is a task that workers perform welding inside the pipe as a confined space. Based on risk assessment, back welding was a high risk task [16]. These authors in another study assessed the welder's exposure to carcinogen metals (Cr, Cd, and Ni). The NIOSH methods were used for sampling and measurement of metals. Back welders group had maximum exposure to total fume and mentioned elements [52].

Determination of occupational exposures to gases must be based on workplace measurements, because the local ventilation and workplace design can affect the actual concentrations of toxic gases (ozone, carbon monoxide, nitrogen oxides) in the welders' breathing zone. Hariri et al. surveyed the appropriate personal sampling methods to measure the welding emissions in small and medium enterprises. They proposed NIOSH methods to evaluate the fumes and direct reading instruments for measurement of gases. Also, they offered some guidelines for correct assessment of welding workplaces [60]. Choonover et al. showed welders were exposed to higher concentrations of NO_2 and O_3 than controls. These gases were collected on pre-treated filters with proper solutions. Then, NO_2 and O_3 were analyzed by spectrophotometry and ion chromatography (IC), respectively [21]. Azari et al. conducted a study to evaluate exposure of mild steel welders to ozone and nitrogen oxides during TIG and MIG welding. OSHA ID214 and NIOSH 6014 methods were used to evaluate ozone and nitrogen oxides, respectively. High exposure of welders to these gases was reported in the study [64]. Golbabaei et al. also used OSHA and NIOSH methods as well as direct reading instruments for sampling and measurement of various gases [65].

Although there are various techniques for monitoring of welding emissions (both fumes and gases) in air samples, selecting the proper ones depends on some parameters. Availability of sampling media, sample storage time, and the simplicity, cost, time and sensitivity of analytical technique are essential to planning proper sampling strategies. It is necessary to consider those workers who probably have the highest exposures due to used materials and processes, the characteristics of their tasks, their postures during welding, the conditions of work environment, and other pollutants from processes in the vicinity of welding environment. It is known that high concentrations of some welding fumes and gases can also be explosive; therefore, the workplace should be tested to ensure a safe working environment [61, 66].

Biological Monitoring - Biological monitoring means the measurement of the concentration of a contaminant, its metabolites or other indicators in the tissues or body fluids of the worker. In some cases, biological monitoring may be a supplementary monitoring for the personal assessment [53]. Another advantage of the biological monitoring is the detection of biological effects of the chemical by monitoring reversible and irreversible biochemical changes. It can be used in the medical treatment to identify the real exposures of chemicals absorbed into the body of employees suspected of over-exposing to a chemical [58]. Airborne contaminants measurement and biological monitoring are complementary procedures used to prevent occupational disease, assess the risk to workers' health, and evaluate the effectiveness of control ways. Biological monitoring must be conducted based on a proper strategy. Careful considerations are required to select the best biological matrix for each component. To obtain valid results, timing sample collection, sample preparation and analytical method used to determine the concentration of components are critical. There are different methods for biological monitoring of some welding emissions. As it is indicated in Table 2, biological media and biological exposure indices (BEIs) have been recommended for some metals and gases emitted by welding processes. Totally, complete information can be provided by biological monitoring and air monitoring to assess the worker exposure to welding emissions.

Ellingsen et al. studied the concentration of manganese in whole blood and urine in welders. Concentration of Mn in whole blood (B-Mn) was about 25% higher in the welders compared to the controls. The increase in B-Mn and the dose-response relation between air-Mn and B-Mn in the welders are strong indicators of Mn. Long-term high exposure to welding fumes may lead to alterations of the urinary excretion of certain cations that are transported through the DMT1 transport system (divalent metal transporter 1 that is found on the surface of the lung epithelial cells) [67]. Kiilunen study showed the metal concentration in post shift urine samples were correlated with the personal air monitoring results. There were statistical significant correlations between urinary concentrations of chromium and nickel and the related total metal concentration in air in wire welding processes. Also, in MIG/MAG welding, chromium is accumulated in the body with a long half life. There is an association between the airborne concentration of nickel and its post shift urinary concentration. In welding, the nickel concentration in post shift urine samples can indicate the body burden [68]. In a study conducted by Hassani et al. the correlation between airborne Mn and urinary Mn was significant for all exposed subjets. The obtained result can introduce the urinary Mn as a biomarker for exposure to this element [69]. Azari et al. measured the serum level of malondialdehyde in welders. Serum MDA of welders was significantly higher than that of the control group. A significant correlation was detected between ozone exposure and level of serum MDA, but the correlation was not observed for nitrogen dioxide exposure [64]. Rossbach recommended the determination of Al in urine for biological monitoring because of the higher sensitivity and robustness of this marker compared to Al in plasma [70]. Golbabaei et al. analyzed the urinary metals among the different groups of welders. According to the results, exposure of welders to fume components leads to more accumulation of them at welders' bodies [52]. Based on different studies, the soluble metal compounds are accumulated in the body, affecting the critical organs. Urinary concentration of metal is used as a biomarker of metal exposure. Therefore, biomonitoring serves as an appropriate tool to monitor both the recent and past exposure and it can be related to the total chemical uptake through all exposure routes [69].

Health monitoring - In addition to the assessment of the airborne concentration of a particular contaminant and its comparison with standard limit, health monitoring may also be done for some hazardous chemicals to assess risks to exposed workers. Health monitoring means monitoring workers exposed to hazardous pollutants to identify changes in their health status and evaluate the effects of exposure. Health monitoring can provide effevtive information to implement proper ways for eliminating or minimizing the risk of exposure and improving control measures. Health monitoring considers all routes of exposure to contaminants [9, 66, 71]. Some tests including spirometry (lung function), audiometry (hearing), biochemical tests (e.g. kidney or liver function), cardiac function tests (heart function), nerve conduction velocity and electromyography tests (nerve and muscle function), and neurobehavioural tests (nerve and brain function) may be used in health monitoring. The type of test used will depend on the occupational hazards that the employee are exposed to [58]. Donaldson [72] and Antonini [73] surveyed lung functions in exposed welders and showed that exposure to welding fumes is associated with both pulmonary and systemic health endpoints, including decrease in pulmonary function, increased airway responsiveness, bronchitis, fibrosis, lung cancer and increased incidence of respiratory infection. In addition to these pulmonary effects, metal fume fever is frequently observed in welders. Exposure to metal fumes and irritating gases cause chronic obstructive pulmonary disease (COPD). Health monitoring of welders can help detect breathing problems and reduced lung functions in early stages, resulting in prevention of further damages. Spirometric tests are used by an occupational phisycian to assess lung functions [74]. Totally, health monitoring may include simple observation of the worker's skin to complicated tests in special cases. Health monitoring must be done by the experienced medical practitioner. An occupational physician can provide specialist services and testing such as spirometric tests, respiratory screening and chest X-rays. It is necessary to do the health monitoring before beginning work with a hazardous chemical to provide enough information for following changes in the worker's health during periods of exposure.

6.2. Risk Assessment of welding emissions

Risk is defined as the possibility of occurance of an event leading to clear concequences. Evaluating risks to workers' safety and health is conducted in risk assessment process. It is performed in some steps including:

- Hazards identification and those at risk
- Evaluating the risks (qualitative or quantitative)
- Elimination or minimization of risks via implementing control measures and taking actions
- Monitoring and reviewing the effectiveness of adopted controls

The severity of hazard and the exopsure level determine the health risk and the type of chemical and nature of work are important factors in this regard. All workers in the vicinity of a special activity should be considered to assess the risk associated with chemical hazards, because they may potentially be at risk of chemicals emmitted by that activity.

In welding environments, employers are resposible to ensure the safety and health of welders and take proper measures for their protection. Although, preventing the occupational risks is the main purpose of risk assessment, it is not possible in all situations; therefore, risks should be reduced using control measures. There are different hazards related to welding process resulting in risks to welders. Chemical hazards, physical hazards, and those associated with ergonomics threaten the health of welders. Since this text deals with air pollution, the risk assessment of welding emissions i.e. fumes and gases is considered. Hazardous chemicals in the workplace result in different risks to workers.

There are different methods to do risk assessment of chemicals in which some principles should be considered. These principles include addressing all relevant hazards and risks and beginning the elimination of risks, if it is possible.

The ministry of manpower of Singapore has published a guideline intitled "semi-quantitave method to assess occupational exposure to harmful chemicals"[75]. This method may be useful to assess the risks resulting from welding emissions. Risk assessment is conducted for following purposes:

- Identifying the hazards related to each harmful chemical
- Evaluating the degree of exposure to chemical of interest
- Determining the likelihood of chemical adverse effects

A risk rating to different tasks can be designate using the mentioned method. After that, using risk rating matrix, hazards are ranked as negligible, low, medium, high and very high (legends 1 to 5) and required actions are prioritized to select appropriate controlling plans. This guideline deals with the health risk to workers exposed to chemicals via inhalation. There are eleven steps for hazard identification and rating, exposure evaluation, and assessing risk. The actual exposure level is required for determination of exposure rating and risk level. A step by step flow chart for assessing the risk, forms needed for completing some steps, and different tables and equations for evaluating the risk have been provided by guideline. All components to assess the risks are available in guideline and it can be used for risk assessment of welding emissions in a simple and fast way. Following, the process flow chart has been presented to understand the consept of risk assessment.

Golbabaie et al. used mentioned guideline to assess the health risks arising from metal fumes on back welders. Risk assessment was performed according to the steps previously explained. Cadmium concentration was ranked as "very high" group. Also, total fumes, total chromium, and nickel were ranked as "high" legend. Findings indicated back welding is a high risk task. High concentration of metals confirmed that working in confined spaces creates a great risk for welders. In some cases as in cadmium despite the rather low concentration of the pollutants, the risk is ranked as "very high" due to the carcinogenisity nature of this element. Therefore, it is not always possible to judge the health hazards of the pollutants based on their concentrations.

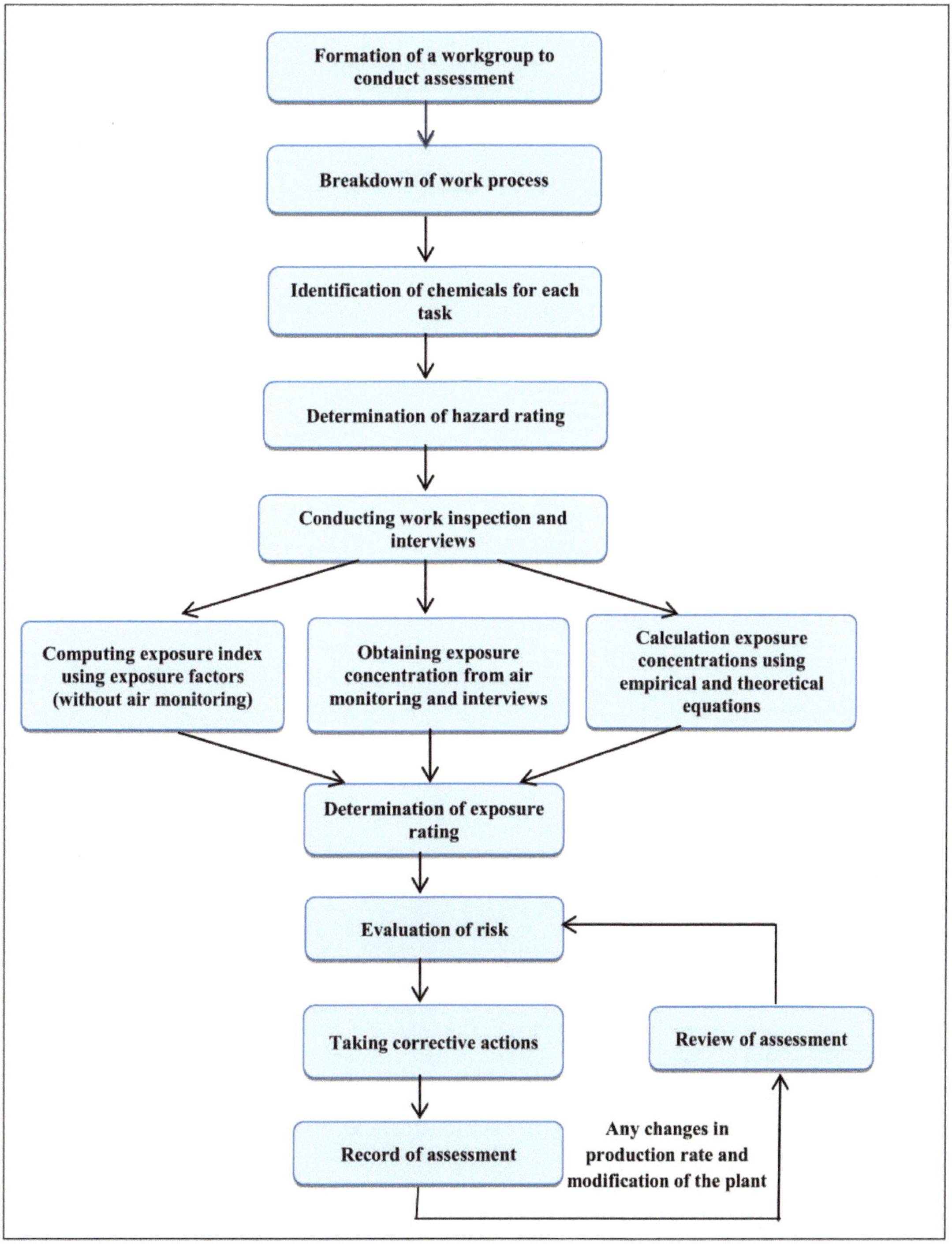

Figure 2. Process flow chart of semi quantitave method for chemicals risk assessment [75]

Following the risk assessment, employers can decide on required preventive measures, the working and production procedures, and also improving the level of welder protection. To complete risk assessment of welding chemicals, data related to air monitoring, biological monitoring, and health monitoring may be required for true judgement. Totally, risk assessment in workplace can result in some advantages. Workers do their tasks in a safe manner; employers provide appropriate programs to prevent high exposure and increase job satisfaction; regulators and related organizations can reliably present health and safety standards. The process of risk assessment is a basis for risk management to reduce welding hazards by choosing correct actions [76-77].

7. Occupational control

Air pollution control deals with the reduction of air pollutants emitted into the atmosphere using different technologies. Sometimes, managing the production process is used to control air pollutant emisstion, therefore, checking the production process can be useful for beginnig the air pollution control. Elimination of a hazard is the first aim to control related risk. In essence, keeping the pollutant emission at the minimum level during the process is the main purpose of controling the air pollution. Based on the risk assessment results, employers can decide for control of risk using proper ways. There are various ways to control the risk of chemicals like welding emissions. If the hazard elimination in not reasonably practicable, other approaches are used to minimize the risk. Substitution, isolation, engineering controls, work practices, and personal protective equipment (PPE) are used to reduce risks to the lowest practicable level in order of priority. Using personal protective equipment is the least recommended control way. To provide a layered safety net, a combination of several control ways may be adopted for preventing risks [66, 76, 78]. In the case of welding, if the elimination of fumes is not practicable, other controling measures should be applied. Modifying the welding process, improving working practices, ventilation, and using PPEs are considered in order to control of fumes.

7.1. Choosing or modifying the welding process

Employers can choose the welding type for production process based upon its efficiency, weld quality, available equipment, and economics. For instance, TIG welding generates less fume compared to MMA, MIG and FCAW processes, so, it can be a proper choice for welding operations. In order to modify the welding process, selecting consumables with minimum fume emissions and considering the welding parameters to minimize the emissions are recommended to employers. The generation of welding fumes is minimized using the lowest acceptable amperage. To optimize the process modification, paying attention to consumables, equipment, and control system is necessary. Selecting proper consumbles leads to minimizing the environmental impacts and controling risks to welders. Welding on non-painted or coated surfaces can also reduce the production of emissions. Process modification in welding results in decreasing needs for administrative controls and other expensive procedures, and also simplifying the process of risk assessment.

7.2. Improvement of working practices

Working practice, the way used to do work, can be improved for control of workers' exposure. Safe work practices are provided by company or organization to perform a task with minimum risk to workforce, environment, and process. Such practices control the manner of performing work and complete engineering measures. Placing the workpiece, as an improving measure, can keep the welders away from plume rising above the weld. Minimizing the welding in confined or enclosed spaces leads to reduction of exposure to pollutants. Proper training programs, housekeeping, maintenance, and doing task on time are the safe welding habits to reduce exposure. Consequently, welding based on safe practices and instructions results in healthier workplace and diminishing the risks of exposure to hazardous emmisions [79, 80].

7.3. Ventilation

Ventilation is the most effective way for removing welding emissions at source to reduce exposure to fumes and gases in welding operations. Designing the ventilation system in accordance with the types of hazardous emissions results in providing a safe atmosphere in the workplace. This control procedure is classified into dilution (general) ventilation and local exhaust ventilation (LEV). The most efficient method to control welding emissions is the combination of LEV and dilution ventilation.

General or Dilution Ventilation -This type of ventilation uses the flow of air into and out of a working environment to dilute contaminants by fresh air. The required fresh air can be supplied by natural or mechanical ways. Dilution ventilation may not be sufficient to control exposure to welding emissions, because it cannot provide enough air movement to prevent the entry of fumes and gases into the welder's breathing zone before removing them from welding environment. In fact, the general ventilation is not suitable for controlling the toxic substances, specially when the worker is downstream of contaminant. To ensure the efficiency of the system, measuring airflow regularly and sampling contaminants to assess exposure are required. A well designed dilution system can be approprite for situations in which welding is done on clean, uncoated, mild steels. In dilution ventilation, draft fans or air-movers, wall fans, roof vents, open doors and windows may be used to move air through the work environment. Totally, if the generated contaminant is in low concentration and can be controlled to the standard exposure level, dilution systems will be effective enough as a control measure [66, 80-82].

Local Exhaust Ventilation - Local exhaust ventilation (LEV), as a primary engineering control, is used to remove contaminants before entering the breathing zone of workers. LEV can be used to control welding emissions close to the generation source. To be effective, LEV system should be well designed and installed, used correctly and properly maintained. Type of generated contaminants and characteristics of the process and work environment are crucial to design LEV [81]. To design a suitable system in welding process, some parameters should be considered, such as fume generation rate, arc- to-breathing zone distance, work practices and worker's exposure. Various parameters related to type of welding have important roles in the fume generation rate and fume composition. Therefore, considering these parameters is necessary to design LEV system [83-85].

For welding processes like stainless steel or plasma arc welding in which fumes containing heavy metals are generated, the LEV system can effectively be used to control worker's exposure. A local exhaust ventilation consists of a hood, fan, duct, and air cleaner. All parts of LEV system must be designed according to correct rules and requirements to remove air pollutants with appropriate efficiency. For instance, the ducting material and structure, air velocity through ducts, the number of branches, and the probability of the leakage and corrosion are important factors related to duct that can affect the LEV system. There are some considereations to select a suitable fan for the system. Some variables such as pressure, flow rate, power, noise, and rotation speed are the main characteristics influencing on the fan performance. Air cleaner is a device to capture welding emissions before it can escape into the ambient air. To select an appropriate air cleaner, some design considereations need to be addressed. Size and shape of welding space, pollutants generation rate, pollutant composition, cost of devices, process type, and the availability of equipment may be effective factors in this respect. In welding processes, source capture systems can be the ideal choise to control fume contaminants using the least air flow rate. In some situations, a source capture system cannot be used. For example situations in which worker has to work on mobile positions; there are a large number of small welding points producing hazardous emissions; welding must be done in confined spaces; and there are some obstructions like overhead cranes leading to problems with ducting installation. Dust collectors (filtration units) and electrostatic precipitators (ESP) can also be used as air cleaners to capture welding emissions before escaping into the environment. ESPs are ideal to collect submicron particles, especially in carbon steel welding. Although the efficiency of ESP is lower than filtration system, it needs very little maintenance and also there is no cost for filter replacement. ESPs are not recommended for stainless steel welding.

Some general considereations should be addressed to design a LEV system. Ducting system should be resistant to the captured emissions; the risks of contaminants accumulation and fire propagation in ducting system should be taken into account; exhausted air containing welding emissions should not be discharged where other workers or people are present; any draught from open doors or windows should be considered because of interference with hood performance. In addition, a maintenance program is required to ensure that control measures remain effective. For instance, regular inspections of LEVsystems should be carried out to check their effectiveness. As an other maintening plan, periodic air monitoring is done to ensure the system has proper performance. Therefore, as well as correct and completed design of LEV system, other elements like employee training, proper use, cleaning, and maintenance are required to achieve the effective protection.

Portable Systems - In some situations, portable systems may be used. These systems are used where welding is infrequently performed and the existing sysrem can be shared between working stations. Also, small mobile units may be used in confined spaces where installing the usual systems is not practical. In these cases, installing the hood close to the emissions point of origin, the hood placement and its distance from the source of welding emissions should be considered. Adequate ventiltion is essential in confined spaces, because the accumulation of hazardous emissions may lead to oxygen deficiency and also adverse effects related to generated fumes and gases. Commercially, there are different portable ventilation systems to use in confined spaces. Flexible air ducts and different kinds of portable fans are available for a variety of ventilation applications. In general, approximately 10 air exchanges per hour should be provided by ventilation in confined spaces. The volume of space and the flow rate of fan determine the time of each exchange. Before entry into the confined space for welding, that space should be ventilated for a minimum of five minutes. It is important to select a proper fan with enough capacity and position it in correct place. Some related organizations have provided procedures and instructions related to working in confined spaces, including ventilation equipment, confined spaces entry, emergency action plan, permit forms, and other requirements for working in these spaces [66, 81, 84, 86].

7.4. Respiratory protection equipments

Personal protective equipment (PPE) should not be used instead of other control measures, but sometimes they may be required along with engineering controls and safe work practices. Respiratory Protection Equipments (RPEs) are used to protect the workers against inhalation of hazardous emissions in the workplace, where exposures cannot adequately be controlled by other ways.

Using a respirator not selected appropriately leads to a false sense of protection for wearer and exposure to hazardous substances. It must be specific to the pollutant and fitted, cleaned, stored and maintained based on provided standards and guidelines for respirators. Each RPE has a protection factor (PF) that is determined as the ratio of the concentration of the pollutant outside the respirator to that inside the respirator. There is a wide range, from low to high, for protection factors. Some organizations like NIOSH have provided required equations and tables to calculate protection factors for respirators. There are different types of respirators and it is possible to select the most appropriate type for existing circumstances. In welding processes, respirators should be selected in accordance with generated emissions, welding type, welding task, and working conditions. For example, NIOSH recommends a self-contained breathing apparatus for welding in confined spaces because the oxygen concentration in the space may be reduced due to welding. Also, a combination of particulate/vapour respirator may be used because of the generation of both of fumes and gases during welding. A standard program is needed for using raspiratory protection devices. Some requirements are followed in this program including hazard assessment, selecting the appropriate respirators in respect of pollutants, respirator fitting test, worker training on how to use respirator correctly, inspection and maintenance of respirator, and recordkeeping. There are two types of RPE. The first type is respirators that clean workplace air before being inhaled and the second type is air-supplied respirators in which air supply is separate from workplace atmosphere. Totally, the suitable RPE for welding processes should be selected by an expert and based on fume concentration, presence of toxic gases, and the probability of oxygen deficiency. Selecting air-purifying respirators with correct filtration cartridge results in protection of welders from low levels of metal fumes and welding gases [87, 88].

8. Conclusion

Air pollution is contamination of the indoor or outdoor environment, leading to changes in the natural characteristics of the atmosphere. In all welding processes, various types of air pollutants are generated. Air pollutants created by welding include fumes and gases whose composition and emission level depend on some factors such as the welding method, welding parameters (current, voltage, shielding gas and shielding gas flow), base metal and other consumables. Exposure to excessive levels of fume and gases can cause different adverse health effects on workers. Since a large number of workers are exposed to welding emissions and also the generated pollutants have negative impacts on environment, a risk assessment program is required to protect workers and environment by suitable procedures. In an effective program, worker's safety and health is considered by management as a fundamental value.Taking different precautions can improve the welder's work situation. There are various techniques for evaluating and monitoring welding pollutants in air samples and biological matrices and also different procedures for their control. Selecting the proper engineering controls can lead to protection of workers and environment. During the risk assessment program and selection of control measures, it is necessary to consider nanoparticles emitted by welding operations. Particle sizes and size distributions of welding emission are critical to determine the efficient control devices. In some cases, breathing zone protection can be used. Health hazards can be reduced by choosing a correct welding helmet and by using the proper shielding gas and welding parameters. It is worth mentioning that proper information should be provided for workers about hazards of their tasks. The welder should be informed of operating techniques and all procedures that reduce welding fumes. The training programs should be included proper ways to perform tasks and proper work practices to reduce fumes. This program includes safety training, monitoring the good safety practices and good environmental practices. Also, the respirator and cartridge selection, fit-testing and respirator maintenance and storage are considered in a suitable training program. Furthermore, employers must be informed about industrial hygiene programs at workplaces and quantitative risk assessment for workers exposed to hazardous compounds. In recent years, different organizations have focused on climate change and environmental impacts of all industrial activities including welding. Various laws, instructions, and guidelines have been provided for protecting the air, environment, and water. Employers are responsible for the purchase of proper welding equipment to meet environmental requirements and choose more environmentally friendly processes.

Author details

Farideh Golbabaei* and Monireh Khadem

*Address all correspondence to: fgolbabaei@sina.tums.ac.ir

Department of Occupational Health Engineering, School of Public Health, Tehran University of Medical Sciences, Tehran, Iran

References

[1] National Institute for Occupational Safety and Health (NIOSH). Nomination of Welding Fumes for Toxicity Studies. In: U.S. Department of Health and Human Services, Public Health Service: NIOSH; 2002.

[2] El-Batanouny MM, Amin Abdou NM, Salem EY, El-Nahas HE. Effect of Exercise on Ventilatory Function in Welders. Egyptian Journal of Bronchology. 2009; 3(1): 103-11.

[3] American Welding Society (AWS). Standard Welding Terms and Definitions. In: Committee TA, 12 ed. Miami: American Welding Society; 2009.

[4] Stern RM. Process-Dependent Risk of Delayed Health Effects for Welders. Environmental Health Perspectives. 1981; 41: 235-53.

[5] Husgafvel-Pursiainen I, Siemiatycki J. IARC Monograph on Welding fumes. 1990; 49.

[6] Erdely A, Antonini JM, Salmen-Muniz R, Liston A, Hulderman T, Simeonova PP, et al. Type I interferon and pattern recognition receptor signaling following particulate matter inhalation. Particle and fibre toxicology. 2012; 9 (1).

[7] Gonser M, Hogan, T., editor. Arc Welding Health Effects, Fume Formation Mechanisms, and Characterization Methods: InTech; 2011.

[8] Carter GJ. Risk Assessment and Control of Exposure During Arc Welding of Steel. Intrernational Conference on Health and Safety in Welding and Allied Processes; 9-11 May; Kopenhagen, Denmark 2005.

[9] Safe Work Australia. Guidance On The Interpretation Of Workplace Exposure Standards For Airborne Contaminants. 2012. p. 1-63.

[10] Finneran A, O'Sullivan L. Force, posture and repetition induced discomfort as a mediator in self-paced cycle time. International Journal of Industrial Ergonomics. 2010; 40 (3): 257-66.

[11] Canadian Centre for Occupational Health and Safety (CCOHS). Welding-Overview of Types and Hazards [Internet]; 2010. Available from: http://www.ccohs.ca/oshanswers/safety haz/

[12] Turan E, Kocal T, unlugencoglu K. Welding technologies in shipbuilding industry. The Online Journal of Science and Technology. 2011; 1 (4).

[13] Occupational Safety and Health Administration. Safety and Health Topic Criteria for a Recommended Standard: Working in Confined Spaces. Washington, DC Occupational Safety & Health Administration; 2007. Available from: www.osha.gov/SLTC/confinedspaces.

[14] American Welding Society (AWS). Confined Spaces, Safety and health Fact Sheet No. 11. 1995.

[15] NIOSH. Hazards to Outdoor Workers [Internet]. Centers for Disease Control and Prevention, Workplace Safety & Health Topics; 2013. Available from: http://www.cdc.gov/niosh/topics/outdoor/.

[16] Golbabaei F, Khadem M, Hosseini M, et.al. Exposure to metal fumes among confined spaces welders. Ital J Occup Environ Hyg. 2012; 3 (4): 196-202.

[17] Weman K. Welding processes handbook. bington Hall, Abington Cambridge, England: Woodhead Publishing Ltd and CRC Press LLC; 2003.

[18] Kopeliovich D. Classification of welding processes: substech; 2012 [updated 2014/09/03]. Available from: http://www.substech.com/dokuwiki/doku.php

[19] Flagan RC, Seinfeld JH. Fundamentals of Air Pollution Engineering. Englewood Cliffs, New Jersey Prentice Hall; 1988.

[20] World Health Organization (WHO). Air Pollution; 2014. Available from: http://www.who.int/topics/air pollution/en/.

[21] Schoonover T, Conroy L, Lacey S, Plavka J. Personal Exposure to Metal Fume, NO_2, and O_3 among Production Welders and Non-welders. Industrial Health. 2011; 49: 63-72.

[22] Amza G, Dobrota D, Dragomir MG, Paise S, Apostolescu Z. Research on Environmental Impact Assessment of Flame Oxyacetylene Welding Processes. METALURGIJA 2013; 52 (4): 457-60.

[23] Nakhla H, Shen JY, Bethea M. Environmental Impacts of Using Welding Gas. The Journal of Technology, Management, and Applied Engineering 2012; 28 (3): 1-11.

[24] Kalliomaki PL, Hyvarinen HK, Aitio A, Lakoma EL, Kalliomaki K. Kinetics of the metal components of intratracheally instilled stainless steel welding fume suspensions in rats. British journal of industrial medicine. 1986; 43 (2): 112-9.

[25] AGA. FACTS ABOUT Fume and gases. AGA (a member of Linde Group), Report No. 110199 0912 – 1.3 HL.

[26] Spear JE. Welding Fume and Gas Exposure. Magnolia, Texas J.E. Spear Consulting, LLC, 2004.

[27] 27. Fuglsang K, Gram LK, Markussen JB, Kristensen JK. Measurement of ultrafine particles in emissions from welding processes. 16th International Conference on Joining of Materials; 10-13 May 2011; Elsinore, Denmark 2011.

[28] Brand P, Klaus L, Uwe R, Kraus T. Number Size Distribution of Fine and Ultrafine Fume Particles From Various Welding Processes. Ann Occup Hyg. 2013; 57 (3): 305-13.

[29] Keane M, Siert A, Stone S, Chen B, Slaven J, Cumpston A, et al. Selecting Processes to Minimize Hexavalent Chromium from Stainless Steel Welding. WELDING JOURNAL. 2012; 91: 241s-246s.

[30] Gomes J. Albuquerque P, Miranda R, Vieira T. On the toxicological effects of airborne nanoparticles from welding processes. IIW European-South American School of Welding and Correlated Processes 18 –20th May 2011, Brazil: 2011.

[31] Gomes J, Miranda R. Emission of airborne ultrafine particles during welding of steel plates. Ciencia & Tecnologia dos Materiais. 2014; 26 (1) :1-8.

[32] Gomes JFP, Albuquerque PCS, Miranda RMM, Vieira MTF. Determination of airborne nanoparticles from welding operations. Journal of Toxicology and Environmental Health, Part A. 2012; 75 (13-15): 747-55.

[33] A-Meo S, Al-Khlaiwi T. Health Hazards of Welding Fumes. Saudi medical journal. 2003; 24: 1-25.

[34] De Flora S. Threshold Mechanisms and Site Specificity in Chromium (VI) Carcinogenesis. Carcinogenesis. 2000; 21 (4): 533-41.

[35] Sowards JW, Ramirez AJ, Lippold JC, Dickinson DW. Characterization Procedure for the Analysis of Arc Welding Fume. Welding Research. 2008; 87: 76-83S.

[36] Keegan GM, Learmonth ID, Case CP. Orthopaedic metals and their potential toxicity in the arthroplasty patient. The Journal of Bone and Joint Surgery. 2007; 89-B (5): 567-73.

[37] pesch B, Haerting J, Ranft U, et.al. Occupational Risk Factor for renal Cell Carcinoma: Agent-Specific Results from a Case Control Study in Germany. International Journal of Epidemiology. 2000; 29: 1014-24.

[38] Erhabor GE, Fatusi S, Obembe OB. Pulmonary functions in ARC-welders in Ile-Ife, Nigeria. Ast Afr Med J. 2001; 78 (9): 461-4.

[39] Tenkate TD. Optical radiation hazards of welding arcs. Rev Environ Health. 1998; 13 (3): 131-46.

[40] Mariutti G, Matzeu M. Measurement of ultraviolet radiation emitted from welding arcs. Health Phys. 1988; 54 (5).

[41] Tenkate TD. Occupational exposure to ultraviolet radiation: a health risk assessment. Rev Environ Health. 1999; 14 (4): 187-209.

[42] Ernst E, Bonde JP. Sex hormones and epididymal sperm parameters in rats following sub-chronic treatment with hexavalent chromium. Hum Exp Toxicol. 1992; 11 (4): 255-8.

[43] Jensen TK, Bonde JP, Joffe M. The influence of occupational exposure on male reproductive function. Occupational Medicine. 2006; 56: 544-53.

[44] Bonde JP. Subfertility in relation to welding. A case referent study among male welders. Dan Med Bull. 1990; 37 (1): 105-8.

[45] Bonde JP, Hanse KS, Levine RJ. Fertility among Danish male welders. Scand J Work Environ Health. 1990; 16 (5): 315-22.

[46] Mortensen JT. Risk for reduced sperm quality among metal workers, withspecial reference to welders. Scand J Work Environ Health. 1988; 14 (1): 27-30.

[47] Sheiner EK, Sheiner E, Hammel RD, Potashnik G, Carel R. Effect of Occupational Exposures on Male Fertility: Literature Review. Industrial Health. 2003; 41: 55-62.

[48] Iregren A. Manganese neurotoxicity in industrial exposures: proof of effects, critical exposure level, and sensitive tests. Neurotoxicology. 1999; 20 (2-3): 315-23.

[49] Bowler RM, Gysens S, Diamond E, Booty A, Hartney C, Roels HA. Neuropsychological Sequelae of Exposre to Welding Fumes in a Group of Occupationally Exposed Men. Int J Hyg Environ Health. 2003; 206: 517- 29.

[50] [50]Wang X, Yang Y, Wang Xi, Xu S. The Effect of Occupational Exposure to Metals on the Nervious System Function in Welders. J Occup Health. 2006; 48: 100-6.

[51] International Agency for Research on Cancer (IARC). welding fumes and gases. lARC Monograhs 49; 1987.

[52] Golbabaei F, Seyedsomea M, Ghahri A, Shirkhanloo H, Khadem M, Hassani H, Sadeghi N, Dinari B. Assessment of Welders Exposure to Carcinogen Metals from Manual Metal Arc Welding in Gas Transmission Pipelines, Iran. Iranian J Publ Health. 2012; 41 (8): 61-70.

[53] National Occupational Health and Safety Commission. Welding: Fumes and Gases. In: Commonwealth of Australia, editor: Ambassador Press Pty Ltd; 1990.

[54] Division of Workers' Compensation. Welding Hazards Safety Program. In: Texas Department of Insurance, editor; 2012.

[55] American Conference of Governmental Industrial Hygienists (ACGIH). Threshold Limit Values for chemical substances and physical agents & biological exposure indices.: ACGIH, Cincinnati, OH; 2010.

[56] Safe Environment-Managing Property Risk. Welding Fume [Internet]. Safe Environment, Sydney; 2012. Available from: http://www.safeenvironments.com.au/welding-fume/.

[57] Ashby HS. Welding Fume in the Workplace, Preventing Potential Health Problems through Proactive Controls. Professional Safety. 2002; Apr: 55-60.

[58] Department of Consumer and Employment Protection. Risk-based health surveillance andbiological monitoring. Resources Safety, Department of Consumer and Employment Protection, Western Australia; 2008.

[59] Pires I, Quintino L, Miranda R, Gomes J. Fume emissions during gas metal arc welding. Toxicological and Environ Chemistry. 2006; 88 (3): 385-94.

[60] Spiegel-Ciobanu V. Occupational health and safety regulations with regard to welding and assessment of the exposure to welding fumes and of their effect. Welding and Cutting. 2012; 11 (1): 61.

[61] Occupational Safety and Health Administration (OSHA). OSHA Technical Manual, Personal Sampling for Air Contaminants In: U.S. Department of Labor, Washington, DC OSHA; 1999.

[62] Education and Information Division. National Institute for Occupational Safety and Health. Occupational Exposure Sampling Strategy Manual. In: Department of Health Education and *Welfare*, Cincinnati: NIOSH; 1977.

[63] Hariri A, Yusof MZM, Leman AM. Sampling Method for Welding Fumes and Toxic Gases in Malaysian Small and Medium Enterprises (SMEs). Energy and Environment Research. 2012; 2 (2): 13-20.

[64] Azari MR, Esmaeilzadeh M, Mehrabi Y, Salehpour S. Monitoring of Occupational Exposure of Mild Steel Welders to Ozone and Nitrogen Oxides. Tanaffos. 2011; 10 (4): 54-9.

[65] Golbabaei F, Hassani H, Ghahri A, Arefian S, Khadem M, Hosseini M, Dinari B. Risk Assessment of Exposure to Gases Released by Welding Processes in Iranian Natural Gas Transmission Pipelines Industry. International Journal of Occupational Hygiene (IJOH). 2012; 4 (1): 6-9.

[66] Safe Work Australia. Managing Risks of Hazardous Chemicals in the Workplace. Australia; 2012.

[67] Ellingsen DG, Dubeikovskay L, Dahl K, et al. Air exposure assessment and biological monitoring of manganese and other major welding fume components in weldersw. J Environ Monit. 2006; 8: 1078-86.

[68] Kiilunen M. Use of biological monitoring for exposure assessment in welding Health and safety in welding and allied processes Brondby, Denmark; 2005.

[69] Hassani H, Golbabaei F, Ghahri A, et al. Occupational Exposure to Manganese-containing Welding Fumes and Pulmonary Function Indices among Natural Gas Transmission Pipeline Welders. J Occup Health. 2012; 54: 316-22.

[70] Rossbach B, Buchta M, Csanady GA, et al. Biological monitoring of welders exposed to aluminium. Toxicology Letters. 2006; 162: 239-45.

[71] Safe Work Australia. Health Monitoring for Exposure to Hazardous Chemicals, Guide for Persons Conducting -A Business or Undertaking. Australia; 2013.

[72] Donaldson K, Tran L, Jimenez LA, Duffin R, et al. Combustion-derived nanoparticles: A review of their toxicology following inhalation exposure. Particle and Fibre Toxicology. 2005; 2 (10): 1-14.

[73] Antonini JM, Lewis AB, Roberts JR, Whaley DA. Pulmonary Effects of Welding Fumes: Review of Worker And Experimental Animal Studies. American Journal of Industrial Medicine. 2003; 43: 350-60.

[74] Health and safety in welding and allied processes - assessing and controlling the exposure risk and complying with the COSHH regulations for welding fume [Internet]. Technical Knowledge 2014. Available from: http://www.twi-global.com/technical-knowledge/faqs/health-and-safety-faqs

[75] Ministry of manpower. A semi-quantitative method to assess occupational exposure to harmful chemical. In: Occupational Health and safety Division, Singapore; 2005.

[76] Europian Commision. Guidance on Risk assessment at Work. In: Directorate General Employment and Social Affairs, Luxembourg: Office for Official Publication; 1996.

[77] National Academy of Sciences. Occupational Health and Safety in the Care and Use of Nonhuman Primates. Washington DC: The National Academies Press; 2003.

[78] Overseas Environmental Cooperation Center. Air Pollution Control Technology Manual. Japan: Environmental Agency, Government of Japan; 1998.

[79] Hewitt PJ. Strategies for risk assessment and control in welding: challenges for developing countries. Annals of Occupational Hygiene. 2001; 45 (4): 295-8.

[80] The welding Institute (TWI). Control of welding fume, Health, safety and accident prevention. [Internet]. The Welding Institute: UK; 2014. Available from: http://www.twi-global.com/technical-knowledge/job-knowledge.

[81] Government of Alberta-Employment and Immigration. Welder's Guide to the Hazards of Welding Gases and Fumes Canada: Workplace Health and Safety Bulletin; 2009.

[82] Health and Safety Executive. General Ventilation in the Workplace, Guidance for Employers. Sheffield: Health and Safety Executive; 2000.

[83] American Welding Society (AWS). Fumes and gases in the welding environment. Miami, FL: AWS; 1979.

[84] Chevron MCBU. Confined Space Entry. USA: HSE, 2012 Contract No. CPL-HES 201

[85] Flynn MR, Susi P. Local exhaust ventilation for the control of welding fumes in the construction industry-a literature review. Annals of occupational hygiene. 2012: 1-13.

[86] OSHA. OSHA and Welding Exhaust In: General Industry Standards and Interpretations, Marine Chemist Service, Inc; 1979.

[87] Health and Safety Authority. A Guide to Respiratory Protective Equipment. Dublin; 2010.

[88] Employment and Social Development Canada. A Guide to Health Hazards and Hazard Control Measures with Respect to Welding and Allied Processes. Canada, Quebec; 2007.

7

Artificial and Biological Particles in Urban Atmosphere

Chang-Jin Ma and Mariko Yamamoto

Additional information is available at the end of the chapter

1. Introduction

In urban areas including smaller towns, the presence of ambient artificial particulate matters (PM) (e.g., aged fine PM and asbestos) and biological particles has been one of the main causes of adverse effects on human health and air quality. Bioaerosols are airborne particles of biological origin (e.g., bacteria, fungi, pollen, viruses). Some biological contaminants trigger allergic reactions, including hypersensitivity pneumonitis, allergic rhinitis, and some types of asthma.

In Japan, there is peculiar annual variation in these ambient PM due to the sources as well as the changing weather. Most synergistic effects take place during spring because of windborne pollen from trees, especially Japanese cedar trees.

From 1949 to 1970, a large number of Japanese cedar trees were planted in many parts of Japanese Island because of a great demand for new housing after World War II. Planted cedar forests now cover 12% of Japan's total land area, which is more than 45,000 km^2. Vast amounts of pollen, especially from cedar trees, drifts over wide areas every spring [1]. As a result, more and more Japanese over the last 30 years have been affected by pollen allergies.

The air quality in urban area, especially its $PM_{2.5}$ (less than 2.5 μm in aerodynamic diameter) level, has become of increasing public concern because of its importance and sensitivity related to health risks [2]. This inhalable $PM_{2.5}$ is generally emitted from both natural (e.g., volcanic eruptions, forest fires, etc.) and anthropogenic sources (combustion processes in industrial sector and automobiles), and it can also be formed when gases react in the air. Inhalable particles, particularly $PM_{2.5}$, have been demonstrated to have the greatest impact on human health, visibility reduction, and solar radiation change, especially in densely populated urban areas.

Asbestos, a naturally occurring fibrous mineral, can be found naturally in the outdoor atmosphere and in some drinkable water, including water from natural sources around the world [3]. As asbestos had been a popular building material since the 1950s, it is still found in many buildings, including hospitals, schools and homes. Inhaling asbestos fibers is known to cause several serious and even fatal lung diseases. Studies have shown that the non-occupationally exposed population have 10,000 - 999,999 asbestos fibers in each gram of dry lung tissue, which translates into millions of fibers and tens of thousands of asbestos bodies in every person's lungs [4]. However, most building material products manufactured today do not contain asbestos. In the industrialized countries, asbestos was phased out of building products mostly in the 1970s, with most of the remainder phased out by the 1980s [5]. In 2006, the Japanese Ministry of Health, Labour and Welfare issued a final ruling banning most asbestos-containing products with the exception of 5 kinds of materials (e.g., some sealing materials). Those 5 unbanned materials were also banned eventually in 2011. Ever since its initial phase out in 2006 and permanent ban in 2011, it can still be found today in some older buildings and consumer goods. Ambient asbestos fibers will finally be lost in the air and eventually precipitate on the ground. These pieces of asbestos are likely to settle on the soil but can be re-released again into the atmosphere [6].

To assess the impact of both artificial and biological PM on the environment, including air quality, ecosystems, and human health, it is necessary to know its concentration, chemical composition, and the interplay among their components. The ambient outdoor PM in urban areas has seldom been evaluated with respect to both artificial and biological components. In light of this situation, we undertook a field campaign to evaluate the artificial and biological PM in an urban environment in Japan during springtime.

2. Material and methods

2.1. Description of Fukuoka prefecture

Fukuoka Prefecture is located on Kyushu Island, Japan's third largest island, located southwest of the main island, Honshu. According to the latest estimates (June 1, 2013), its population and total area are 5,088,480 and 4,971 km^2, respectively [7].

Fukuoka Prefecture faces the sea on three sides, bordering on Saga, Oita, and Kumamoto prefectures and facing Yamaguchi Prefecture across the Kanmon Straits. Fukuoka Prefecture includes the two largest cities on Kyushu, Fukuoka and Kitakyushu, and much of Kyushu's industry. Fukuoka prefecture's main cities form one of Japan's main industrial centers, accounting for nearly 40% of the economy of Kyushu. Major industries include automobiles, transport equipment, electronic parts and machine, general machinery, iron and steel, semiconductors, steel, and food [8].

2.2. Sampling and monitoring of ambient PM

An intensive measurement of PM was conducted at four selected sites (A: 33.65°N; 130.45°E, B: 33.63°N; 130.42°E, C: 33.39°N; 130.26°E, D: 33.66°N; 130.40°E,) in the Fukuoka Prefecture. The location of each sampling site in Figure 1 is indicated by filled circles.

Site A has six-lane roads with heavy traffic conditions. Site B is a residential area and surrounded by a residential area without any known point sources. Site C is an industrial area with roughly 60 manufacturing companies including small-to-medium sized enterprises. Most of them are distributed within a radius of one kilometer from site C. Site D is a desolate area with resort beach without the influx of people, as our study was conducted during an off-season period.

An intensive measurement of PM was simultaneously conducted at four-sites for two days beginning on April 18, 2007 and ending in April, 2007. For the sampling of size-classified PM, four impactor samplers (Tokyo Dylec Co.) were simultaneously operated. Particles were collected directly on the filter arranged behind the jet-nozzles of the 1st stage of sampler. For the collection of particle samples, airflow was maintained at 20 l min^{-1}. The size fraction of PM and filter kind at each stage are coarse fraction (> 2.5 μm) at the first stage (a 47 mm diameter, non hole Nuclepore® polycarbonate filter) and fine fraction ($PM_{2.5}$) at a back-up stage (a 47 mm diameter, 0.01 μm hole Nuclepore® polycarbonate filter), respectively.

Although the influx value (i.e., deposition rate, grains cm^{-2}) is a widely used method for studies of airborne pollen, in this study, for the purpose of measuring the ambient concentration (grains m^{-3}), airborne pollen grains were also collected on the 1st stage of impactor samplers. The directly collected pollen grains on the natural filter surface are more easily observed by a SEM without filter pretreatment compared to those collected on a traditional plastic tape coated by adhesive.

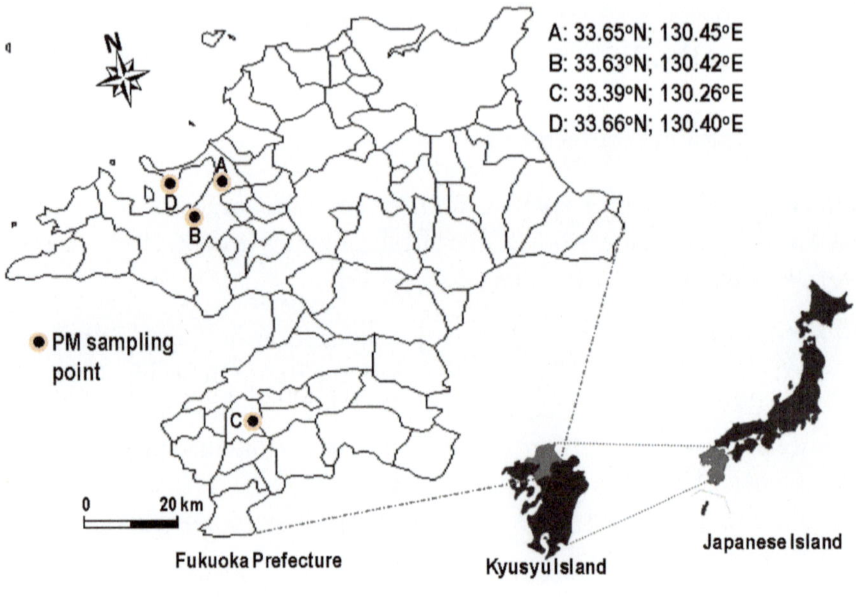

Figure 1. Locations of sampling stations for PM in Fukuoka Prefecture

In order to assess $PM_{2.5}$ mass concentration, 4 Dust scan Scouts (Rupprecht & Patashnick Co. Model 3020) were simultaneously operated at each site. This $PM_{2.5}$ monitoring system makes use of near-forward light scattering to measure the concentration of particulate matter in ambient air. The light source is a safety-interlocked laser that operates at a wave length of 670 nm. The scattered light caused by the presence of particles is received by a sensor, forming the basis of the monitor's computations. During sampling period, there was no Asian dust event. The wind speed was measured at 2.4 - 4.7 m s^{-1}, with the temperature at around 11.2 - 19.4 °C and the average relative humidity at 56%.

2.3. Sample pretreatment and analysis

2.3.1. Sample pretreatment

Figure 2 schematically illustrates the procedures of sample pretreatment and analysis. After sampling, filters were placed in sterilized airtight petridishes and stored in a refrigerator until analysis.

For the laboratory analysis, the filters capturing $PM_{2.5}$ were extracted with deionized water by ultrasonic treatment. And then the extracted water was filtrated through a 25 mm diameter Nuclepore® filter with 0.08 μm pore size to separate into the soluble and insoluble fractions. After filtration, the filtrate was considered to be soluble fractions. Blank filters were handled in the same manner as the samples.

Meanwhile, the coarse particles (>2.5 μm) deposited on the first stage were progressed to single particle analysis for identification of asbestos and pollen. Samples for observation of pollen were coated with a very thin layer of platinum by a machine called a sputter coater.

2.3.2. Analysis of major ionic species in $PM_{2.5}$

In general, sulfate, nitrate, and ammonia had the greatest ambient concentrations in particles [9,10]. The concentrations of major ions in $PM_{2.5}$ (ammonium, nitrate, and sulfate) were determined by Ion Chromatography (IC) (Dionex DX-100). To attain the reliability of the analyzed data, the quality assurance and quality control (QA/QC) was conducted by analyzing a set of known standard species. The data obtained by 5-time repeated IC analyses were tested for precision by checking the relative standard deviation (% RSD, (SD / mean) x 100) of each concentration in standard solution (0.5, 2, and 5 mg/L). As a result, all three ionic species maintained low % RSD levels (i.e., ammonium: 4.63 - 7.59%, nitrate: 0.04 - 9.03%, and sulfate: 1.96 - 9.72%). This high reproducibility is a clear indication of a methodological soundness.

2.3.3. Identification of asbestos and pollen

The most common method of identifying asbestos fibers in ambient PM is by polarized light microscopy. However, a Scanning Electron Microscopy (SEM) can be also usefully applied to the observation of asbestos. The advantage of using a SEM for asbestos identification is that it has better resolution through higher magnification and a greater depth of focus than polarized light microscopy. Most of all, a SEM equipped with an energy dispersive X-ray spectrometer (EDX) (i.e., SEM-EDX) can quantify the elemental components in asbestos. In Figure 3, the diagram of a SEM-EDX is schematically illustrated.

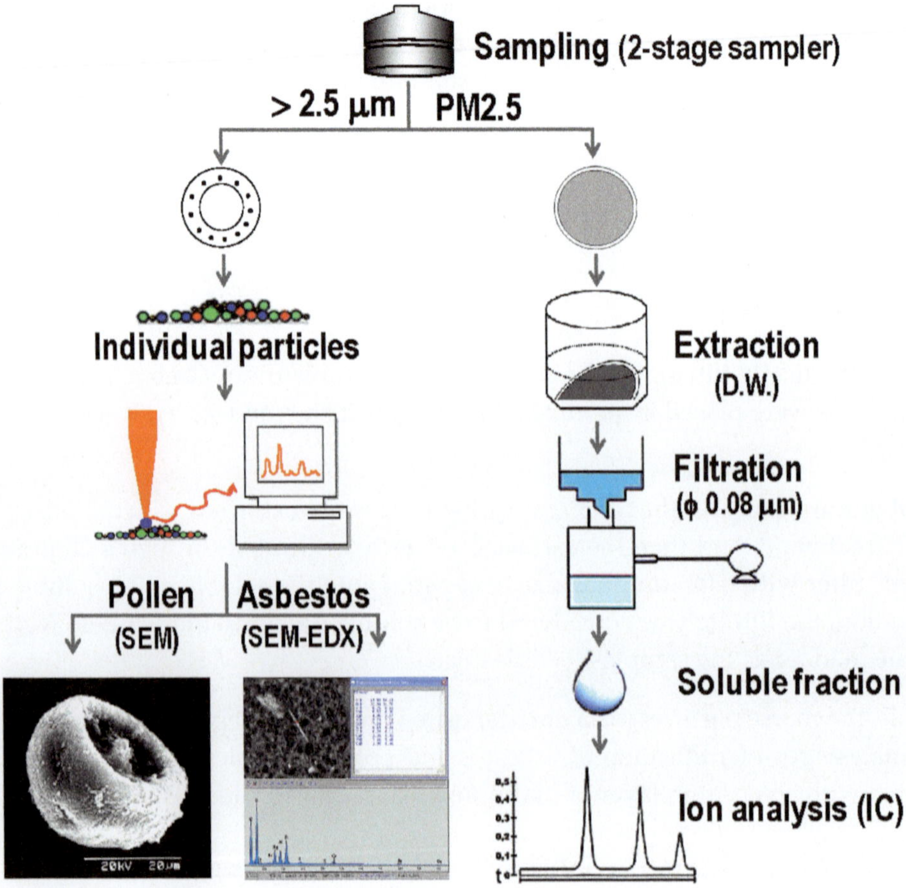

Figure 2. Procedures of sample pretreatment and analysis

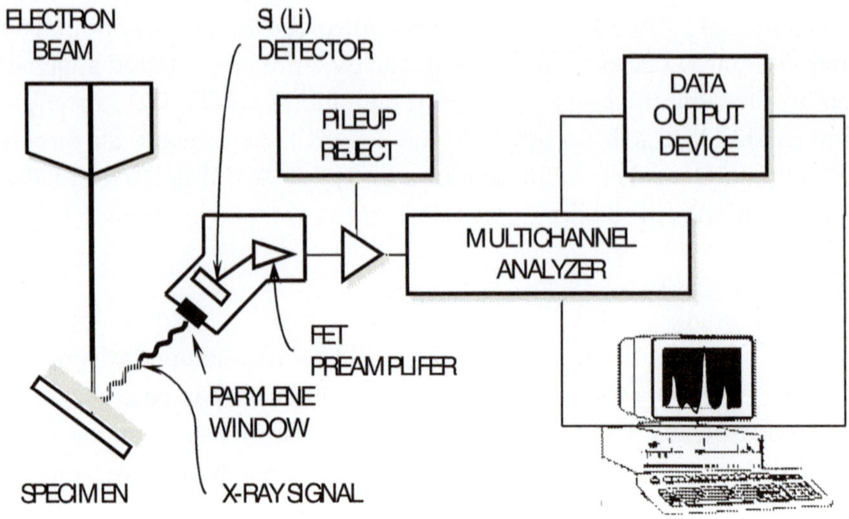

Figure 3. Schematic diagram of a SEM-EDX

Using an EDX and a computer system, information about the elemental properties of asbestos fibers can be gathered and graphed in their appropriate relative ratios. Computation of the exact ratios of the elemental compositions in asbestos fiber allows the researcher to distinguish not only one type of asbestos fiber from another but also asbestos fibers from non-asbestos fibers.

In this study, for the purpose of observing and analyzing the morphological and chemical properties of airborne asbestos and pollen grains, an SEM (JEOL JSM-5400) equipped with an EDX (Philips, EDAX DX-4) was employed. The samples were placed inside the SEM's vacuum column (10^{-6} Torr) through an air-tight door. Pollen species were also distinguished and countered under 3000 x magnification and 15 - 20 kV working conditions.

Calculation of the airborne asbestos fiber concentration on the filter sample was carried out using the following formula:

$$F_{con} = \frac{A_{eff} \times N_{total}}{a_{field} \times n_{field} \times V_{air}}$$

where F_{con} is airborne fiber concentration (fiber/L, f L^{-1}), A_{eff} is effective collecting area of filter (cm^2), N_{total} is the total number of fibers in an SEM field area, a_{field} is an SEM field area (cm^2), n_{field} is total number of fields counted on the filter, and V_{air} is total air volume (L) calculated by sample collection time (min) and pump flow rate (L/min).

3. Results and discussion

3.1. Spatial distribution of $PM_{2.5}$ and major water-soluble ions

Air pollution, especially $PM_{2.5}$, has become a serious issue in East Asia, and there is rising public criticism regarding its effects. Concern in Japan is also increasing as winds transfer the pollution into domestic areas. Figure 4 shows the distribution of $PM_{2.5}$ monitoring data at four-sampling sites (i.e., a heavy traffic area (A), a residential area (B), an industrial area (C), and a desolate area (D)).

Site C, an industrial area, showed the heighest $PM_{2.5}$ (65.3 μg m^{-3}) followed by site A (35.3 μg m^{-3}), site B (22.0 μg m^{-3}), and site C (11.3 μg m^{-3}). As might be expected, the highest $PM_{2.5}$ level was monitored at site C where there is a compact mass of manufacturing companies. Among the four monitoring sites, two sites (i.e., site A and C) exceeded the Japan's $PM_{2.5}$ criteria (a daily average of 35 μg m^{-3}). The reason for the relatively high $PM_{2.5}$ in both heavy traffic and industrial areas might be that much of fine particles and their precursors came from the vehicle emissions and fuel combustion and manufacturing processes in factories. The $PM_{2.5}$ data monitored in this study are compared with those measured in other urban and rural areas in Asia during the springtime. Zhang *et al.* [12] reported that $PM_{2.5}$ was 145.3 μg m^{-3} in Beijing during a non-Asian dust period in the springtime. Meanwhile, in Gosan, a typical rural area in Korea, $PM_{2.5}$ during non-Asian dust period was measured at 26.1 μg m^{-3} [13].

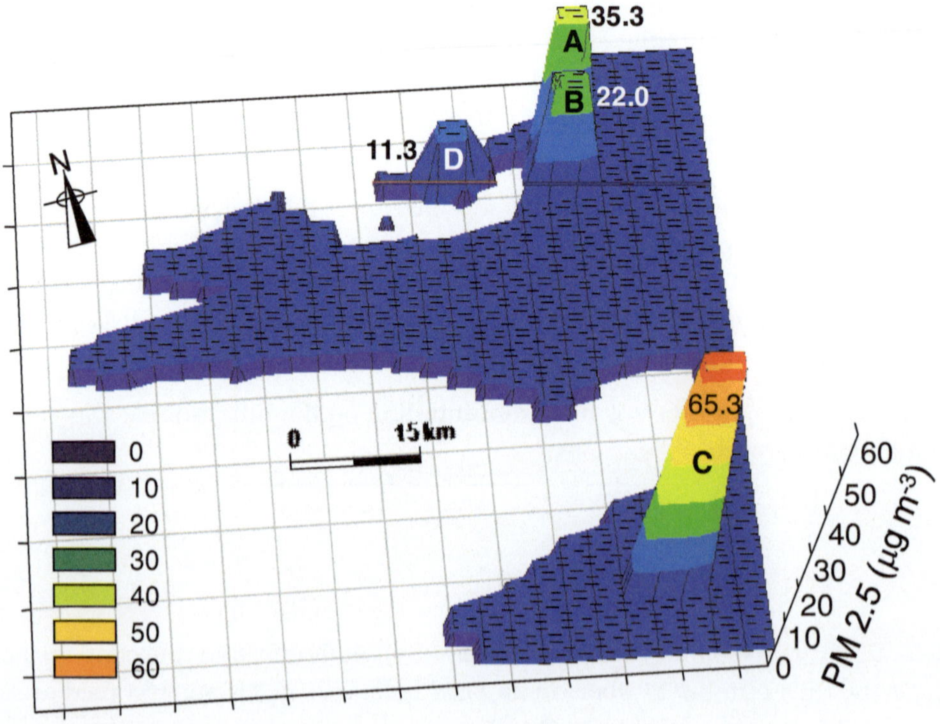

Figure 4. distribution of PM $_{2.5}$ monitoring data at 4-sampling site

It is a matter of course that a large number of factories contributed to the high $PM_{2.5}$ level and caused the regional worsening air pollution in a residential area (B). However, although the "yellow dust" warning did not issue during our field measurement, there is a possibility of the long-range transport of $PM_{2.5}$ and its precursors from the Asian continent to the local study site in Japan. Thus, it is required to clear the uncertainties regarding the linkage between locally high $PM_{2.5}$ at site C and its long-range transport under the springtime meteological conditions. In this study, in order to determine the source region of aerosols at the site C, the atmospheric backward dispersion model was applied.

Figure 5 displays the backward aerosol dispersion simulated by the NOAA Air Resources Laboratory (ARL) HYSPLIT dispersion-trajectory model started from site C. The area scales mean the integrated mass concentration [mass m^{-3}] at 100 - 1000 m height of site C (33.39°N; 130.26°E).

A detailed model description of HYSPLIT was given in reference [14]. According to the result of the HYSPLIT model, a high value of aerosol concentration at the present receptor (measurement location) was not driven from the Chinese continent but was generated from a local area.

Ambient concentrations of major ionic species (i.e., nitrate, sulfate, and ammonium) associated with $PM_{2.5}$ collected at four-sampling sites were overlapped with a map of Fukuoka Prefecture (Figure 6). As shown in Figure 6, the concentrations of major ionic species turned out to be of

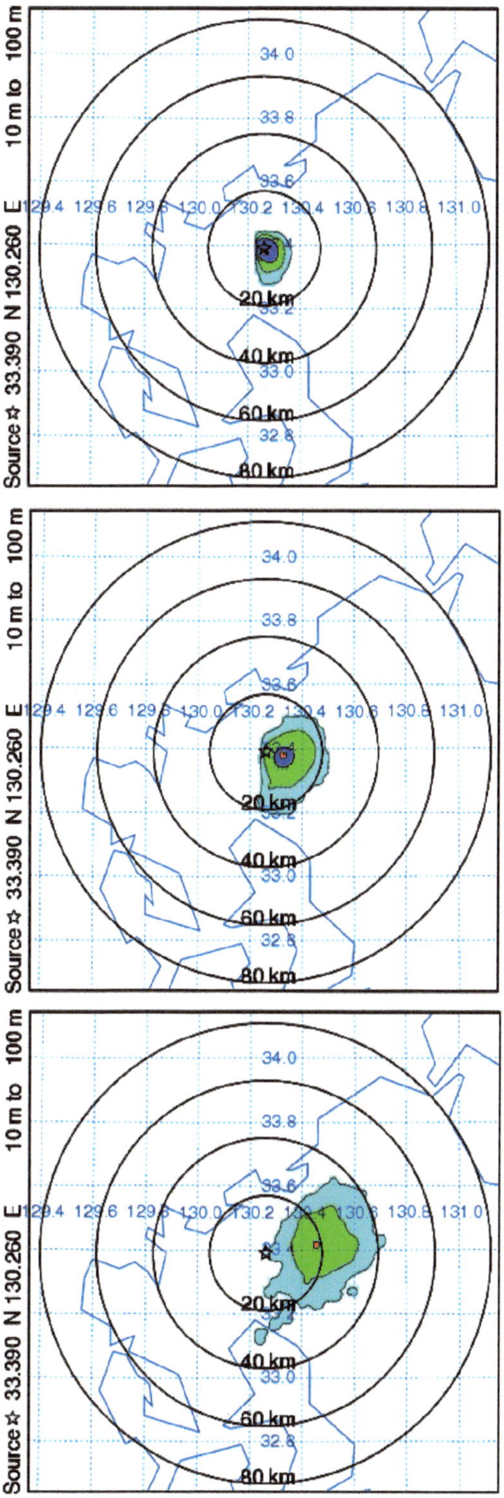

Figure 5. Backward aerosol concentrations simulated by the NOAA ARL HYSPLIT model. The area scales mean the integrated mass concentration [mass m-3] at 10-100 m height of site C (33.39oN; 130.26oE)

considerable variation among four sites. Sulfate was the most abundant species to record the highest concentrations in all urban areas of Fukuoka Prefecture (a heavy traffic area (14.1 μg m^{-3}), an industrial area (30.5 μg m^{-3}), and a residential area (2.8 μg m^{-3}). The sum concentrations of NH_4^+, NO_3^-, and SO_4^{2-} varied in a similar way, as $PM_{2.5}$ (i.e., site C (38.4 μg m^{-3}) > site A (18.3 μg m^{-3}) > site D (5.4 μg m^{-3}) > site B (4.9 μg m^{-3})).

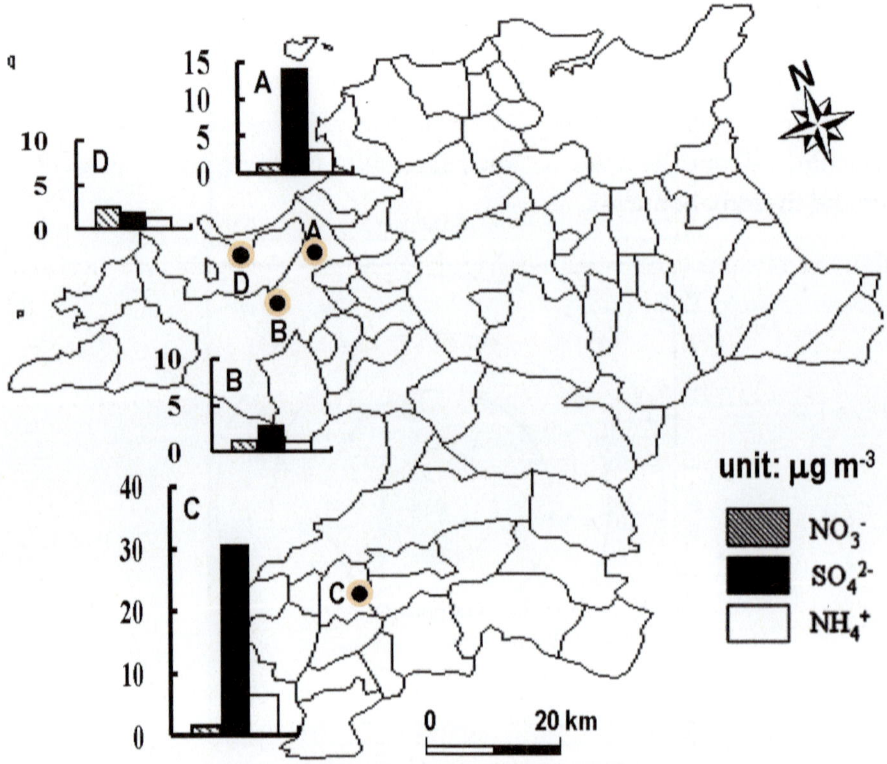

Figure 6. Ambient concentrations of major water-soluble ions associated with $PM_{2.5}$ collected at 4-sampling site in Fukuoka Prefecture

This suggests that the overwhelmingly high level of site C should be associated with the local pollution emissions. In the case of site C and A, the sum of three ionic components correspond to 58.8% and 51.8% of $PM_{2.5}$, respectively. Therefore, it seems reasonable to say that $PM_{2.5}$ in sites C and A was mainly composed by the secondary inorganic aerosol, which was formed by a gas-to-gas reaction in the atmosphere.

The concentrations of three kinds of ionic species in $PM_{2.5}$ in this study are comparable to those of urban areas in Beijing, China, and Durg, India. Gao *et al.* [10] reported that the concentrations of NH_4^+, NO_3^-, and SO_4^{2-} in $PM_{2.5}$ collected in Beijing were 20.5, 15.2, and 42.3 μg $m^{-3,}$ respectively. On the other hand, those concentrations in Durg, a heavy traffic and industrial area, were marked as 2.1, 3.16, and 6.75 μg m^{-3}, respectively [9]. The occupation ratio of the sum concentration of three ionic components in this area (0.89%) is greatly dissimilar to those of this study.

3.2. Assessment of airborne asbestos fiber

Even though asbestos is closely regulated in the present, the deposited and accumulated asbestos fibers for the past several tens of years can be resuspended from the ground near the distributing and manufacturing shops of asbestos products.

Asbestos from natural geologic deposits is known as "naturally occurring asbestos" (NOA) [14]. Health risks associated with exposure to NOA are not yet fully understood. As air quality associated with asbestos in ambient outdoor air has seldom been evaluated in Fukuoka Prefecture, we evaluated airborne asbestos in the urban environment there. Six mineral types are defined by the United States Environmental Protection Agency as asbestos [15]. Among them, crocidolite and amosite are commonly known to cause negative health effects such as lung cancer and mesothelioma [4].

Figure 7 shows an example of asbestos image of SEM, the elemental wt% list, and elemental spectrum obtained from EDX analysis.

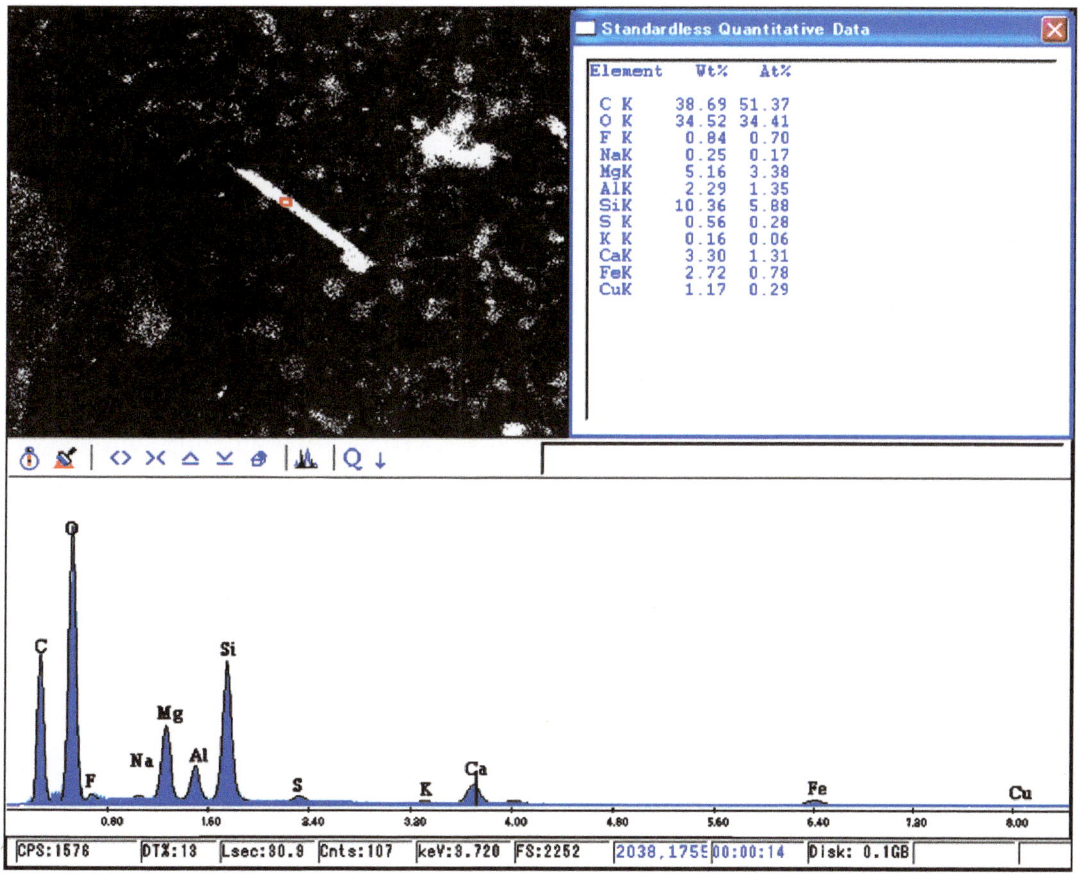

Figure 7. An example of asbestos image of SEM (left upper) and the elemental wt% list (right upper) and spectrum (bottom) obtained from EDX analysis

A part of results of SEM-EDX analysis for asbestos fibers and the processes of classifying asbestos types based on the SEM-EDX elemental wt% data are illustrated in Figure 8.

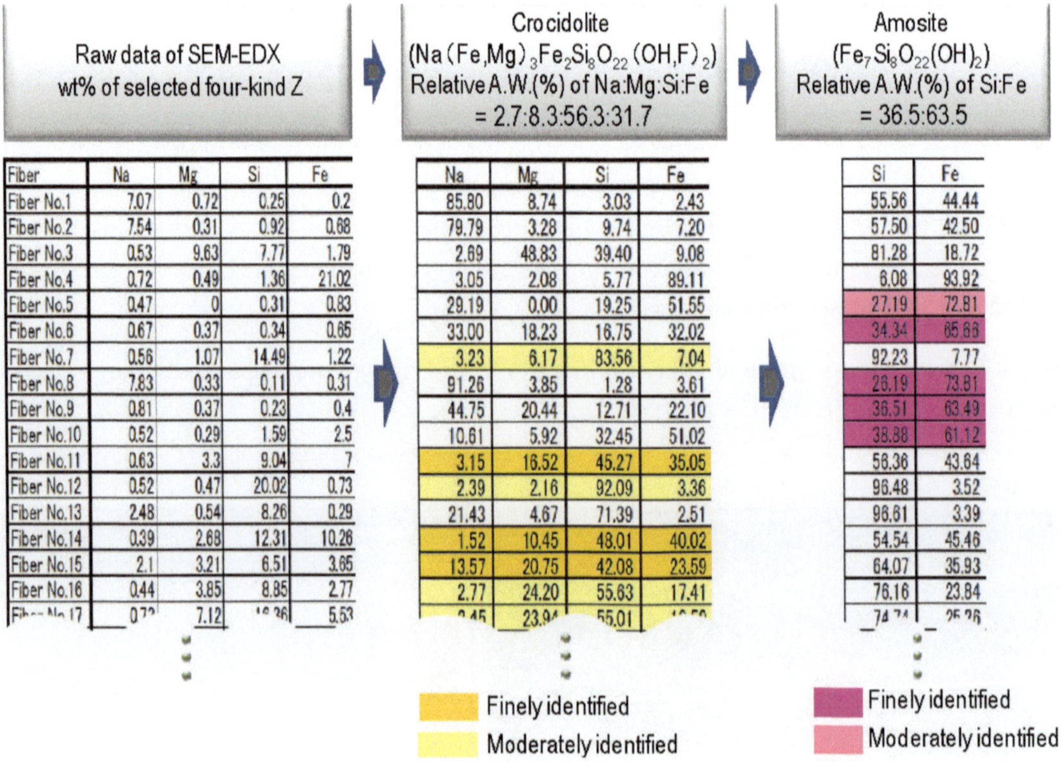

Fiber	Na	Mg	Si	Fe	Na	Mg	Si	Fe	Si	Fe
Fiber No.1	7.07	0.72	0.25	0.2	85.80	8.74	3.03	2.43	55.56	44.44
Fiber No.2	7.54	0.31	0.92	0.68	79.79	3.28	9.74	7.20	57.50	42.50
Fiber No.3	0.53	9.63	7.77	1.79	2.69	48.83	39.40	9.08	81.28	18.72
Fiber No.4	0.72	0.49	1.36	21.02	3.05	2.08	5.77	89.11	6.08	93.92
Fiber No.5	0.47	0	0.31	0.83	29.19	0.00	19.25	51.55	27.19	72.81
Fiber No.6	0.67	0.37	0.34	0.65	33.00	18.23	16.75	32.02	34.34	65.66
Fiber No.7	0.56	1.07	14.49	1.22	3.23	6.17	83.56	7.04	92.23	7.77
Fiber No.8	7.83	0.33	0.11	0.31	91.26	3.85	1.28	3.61	26.19	73.81
Fiber No.9	0.81	0.37	0.23	0.4	44.75	20.44	12.71	22.10	36.51	63.49
Fiber No.10	0.52	0.29	1.59	2.5	10.61	5.92	32.45	51.02	38.88	61.12
Fiber No.11	0.63	3.3	9.04	7	3.15	16.52	45.27	35.05	56.36	43.64
Fiber No.12	0.52	0.47	20.02	0.73	2.39	2.16	92.09	3.36	96.48	3.52
Fiber No.13	2.48	0.54	8.26	0.29	21.43	4.67	71.39	2.51	96.61	3.39
Fiber No.14	0.39	2.68	12.31	10.26	1.52	10.45	48.01	40.02	54.54	45.46
Fiber No.15	2.1	3.21	6.51	3.65	13.57	20.75	42.08	23.59	64.07	35.93
Fiber No.16	0.44	3.85	8.85	2.77	2.77	24.20	55.63	17.41	76.16	23.84
Fiber No.17	[illegible]	7.12	[illegible]	5.53	[illegible]	23.94	55.01	[illegible]	74.74	25.26

Figure 8. The raw data of SEM-EDX analysis for several asbestos fibers and the processes of classifying asbestos types based on the SEM-EDX elemental wt% data are illustrated in

Figure 9 illustrates the number concentration of asbestos fiber (a sum of crocidolite (Na(Fe,Mg)$_3$Fe$_2$Si$_8$O$_{22}$(OH,F)$_2$) and amosite (Fe$_7$Si$_8$O$_{22}$(OH)$_2$) at four-sampling sites and distribution of point sources of asbestos fiber in Fukuoka Prefecture. The highest airborne asbestos fiber concentration was recorded at site C (14.4 f L^{-1}), followed by site A (5.9 f L^{-1}), site D (3.4 f L^{-1}), and site B (2.5 f L^{-1}). The average concentration of the airborne asbestos fiber at all sites was 6.14 f L^{-1}. This average concentration level is similar to those (7 f L^{-1}) measured at asbestos abatement sites in Korea [16].

In 1989, the Japanese Air Pollution Control Law and related orders were revised to classify asbestos as a "specified dust" and to set up 10 fibers/liter (for including all type of asbestos) as the regulation guideline Concentration at the Boundary of the Asbestos Dusts Generation Facilities (i.e. asbestos products manufacturing facilities) [6]. Asbestos fiber concentration of site C (14.4 f L^{-1}) is considerably higher than the regulated levels of asbestos of the Concentration Standard at the Boundary of the Asbestos Dusts Generation Facilities.

To explore the possible reasons for the high asbestos fiber concentration at site C, the forecast atmospheric dispersion was simulated by the NOAA ARL Gaussian model. Figure 10 shows

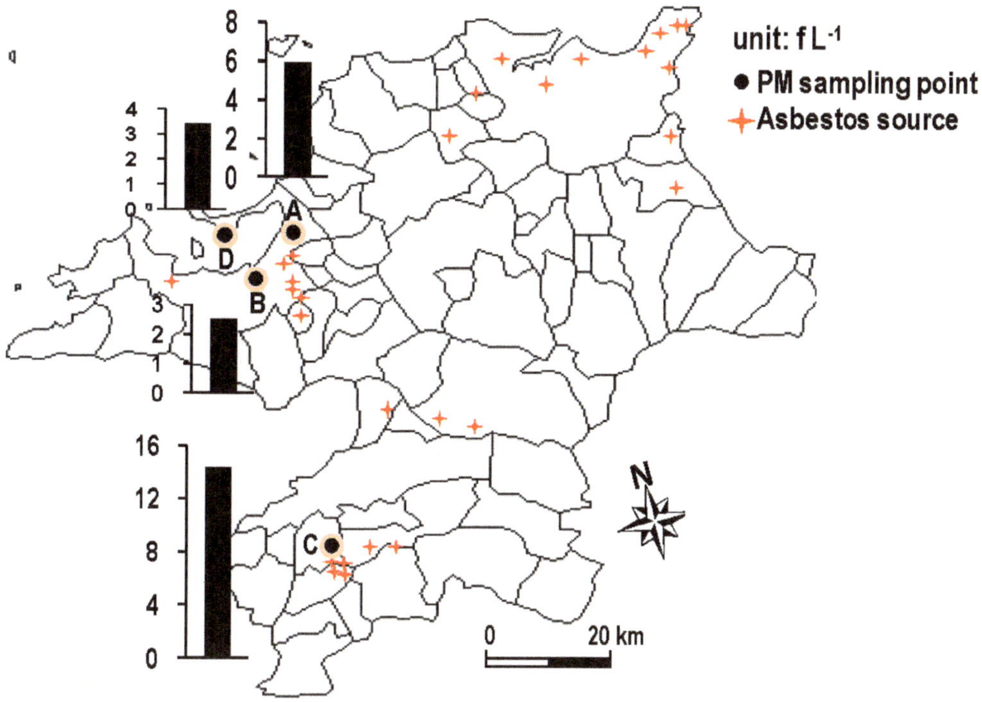

Figure 9. Number concentration of asbestos fiber (crocidolite and amosite) at 4-sampling site and distribution of point sources of asbestos fiber in Fukuoka prefecture.

sampling site C in Figure 9 having a dense asbestos point source and the area distribution of the air flume started from intensively distributed companies producing asbestos related products.

According to the result of NOAA's ARL Gaussian model, the C sampling site was in an downwind position. It is therefore suggested that the relative high asbestos concentration at site C was strongly influenced by the clumped distribution of point sources of asbestos fiber. Meanwhile, although the concentration of asbestos measured at site A (heavy traffic area) was lower than the national Concentration Standard at the Boundary of the Asbestos Dusts Generation Facilities, asbestos was detected at levels around 2 times higher than those of sites B and D.

NOAA's ARL Gaussian model shown in Figure 10 indicates that the point sources of asbestos fiber situated densely near site A did not exert a direct influence on the asbestos concentration at site A. For several years, automobile parts that needed insulation from heat and friction were manufactured from dangerous asbestos, due to its excellent heat-resistant qualities. Such parts included brake linings, clutch facings, transmission components, disc brake pads, drum brake linings, and brake blocks [5].

According to both the result of NOAA's ARL Gaussian model (Figure 10) and the information on extensive use of asbestos in automobiles, the asbestos concentration at site A was probably affected by automobiles.

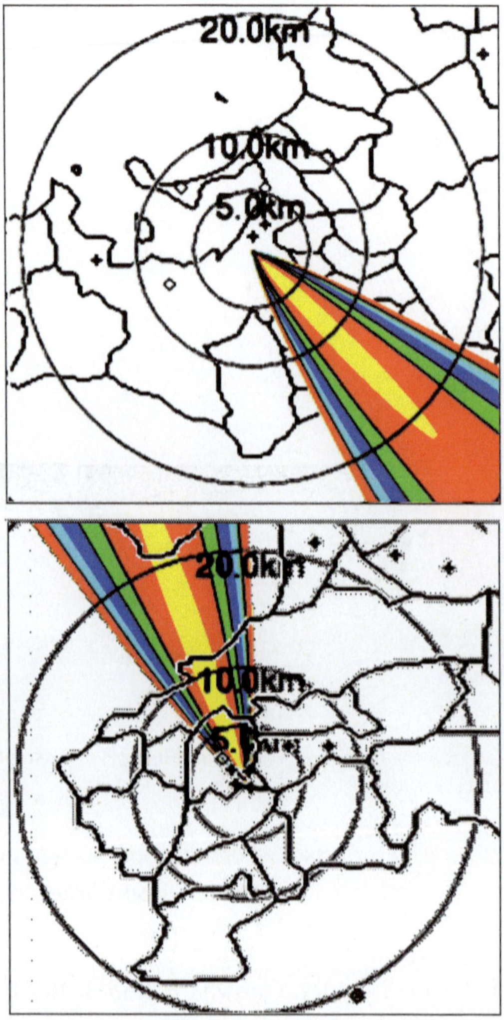

Figure 10. Map showing sampling site C in Figure 9 having dense asbestos point source and the area distribution of air flume started from the intensively distributed companies producing asbestos

3.3. Spatial distribution of ambient pollen

It can be considered that pollen contributes to the organic carbon fraction in totoal suspended particles (TSP). In addition, some plant types can produce pollen in huge quantities. For example, a single ragweed plant can generate a million grains of pollen a day. Therefore, though pollen granules are small and light, they can also attribute to the mass concentration of ambient aerosol particles during the main pollen season [17]. From this point of view, to study the species of pollens, their distribution and concentration can be helpful to understand their ambient behavior and public health, etc.

SEM images of several types of pollens are shown in Figure 11. Classification of pollen grains using SEM was based on morphological characteristics such as shape, size, apertures and ornamentation. The combination of these characteristics makes some pollen grains easily identifiable. However there were also pollen grains sharing several common characteristics that makes identification difficult.

Figure 12 shows the spatial variation of the number concentration of three-kind of airborne pollen at the four sites in Fukuoka Prefecture. Three-type mainly identified pollen grains are reported here. There was a noticeable spatial difference in the concentrations among three pollen types. Cedar (Cryptomeria, also called the Sugi tree in Japanese) pollen was distributed in advance of other pollens at sites A, B, and C, with the maximum concentration at site A. This cedar pollen is the most common allergen for seasonal allergic rhinitis in Japan. Cedar showed a higher concentration than other types of pollen and it was probably generated from the Sugi tree that is the most important timber tree in Japan.

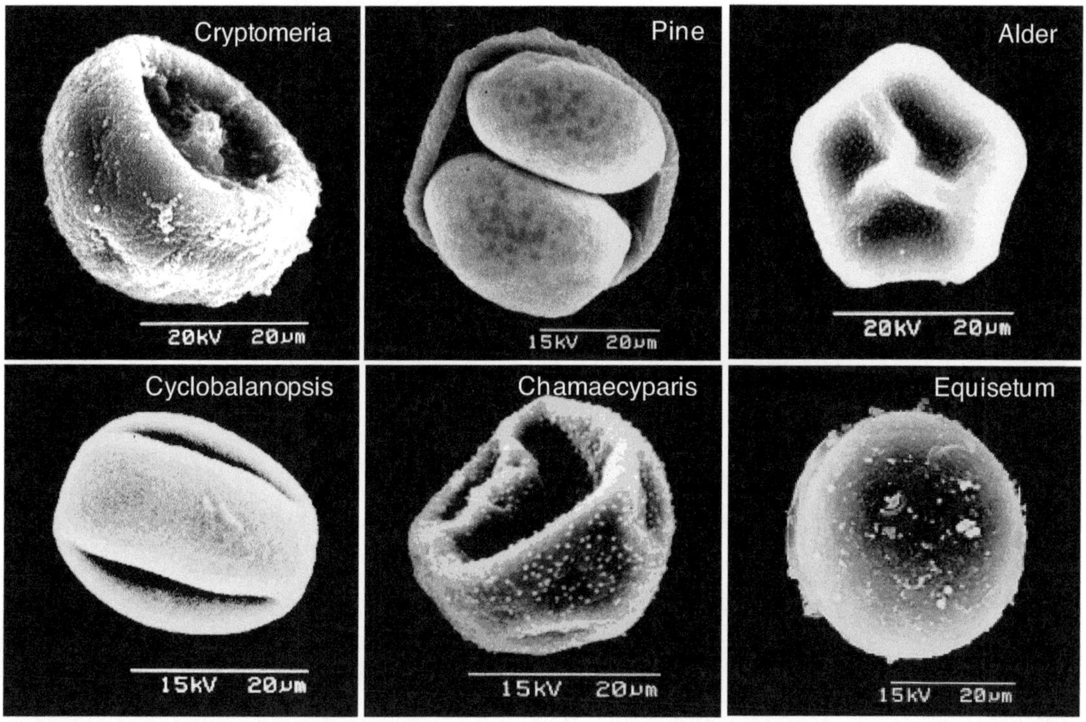

Figure 11. Morphologies of several types of pollen grains (Cryptomeria, Pine, Alder, Cyclobalanopsis, Chamaecyparis, and Equisetum) identified by SEM observation

The sum of the number concentration of three pollen types is still high at site A. Although the atmospheric presence of pollen grains is likely to vary depending on the local kind of plant, actual weather situations, and pollinating period, the vegetation of the surface is also important [18]. Cedar pollen easily absorbs water in the atmosphere and then settles down on surface. However, they can be easily resuspended into the atmosphere by the urban surface covered with asphalt and cement. Both hot-island and building wind can also promote the resuspension of pollen [19]. Therefore, site A, with its relatively high pollen concentrations, was probably affected by the typical urban surface and local plant.

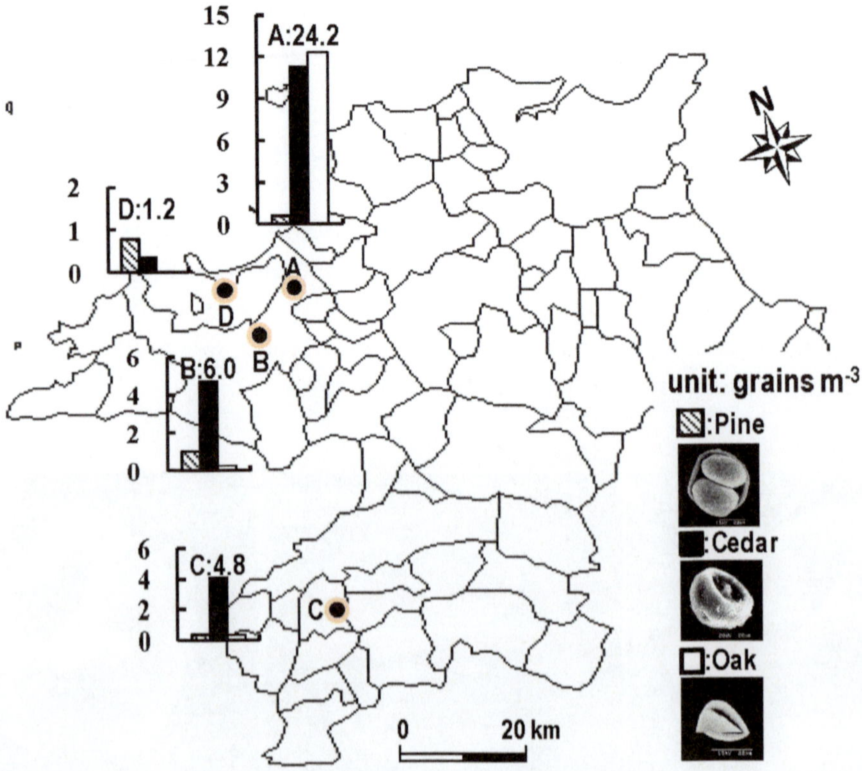

Figure 12. Pollens number concentration at each sampling site

4. Conclusion

A better knowledge of the impact of both artificial and biological PM on urban atmosphere can help establish improved the management strategies of urban air quality. This study focused on a comprehensive and detailed interpretation for the springtime air quality influenced by both artificial (particulate matter (PM) and asbestos) and biological (pollen) sources in Fukuoka Prefecture, Japan. An intensive measurement of PM was conducted at four characteristic sites (i.e., a heavy traffic area, a residential area, an industrial area, and a desolate area) in the Fukuoka Prefecture during spring of 2007. Analysis of major ionic species in $PM_{2.5}$ was performed by an Ion Chromatography, and asbestos and pollen were identified by Scanning Electron Microscopy with an energy dispersive X-ray spectrometer (EDX). $PM_{2.5}$ concentration (65.3 μg m^{-3}) measured in an industrial area (site C) was extraordinarily high compared to those monitored in other areas; it greatly exceeded the Japan's $PM_{2.5}$ criteria (a daily average of 35 μg m^{-3}). NOAA's HYSPLIT dispersion model suggests that this high level of $PM_{2.5}$ monitored at site C is unlikely to affect the Asian continent. The ambient concentrations of $PM_{2.5}$-related anions (NH_4^+, NO_3^-, and SO_4^{2-}) and their relative contributions to $PM_{2.5}$ were also investigated in four study areas. The concentrations of these major water-soluble ions exhibit not only strong spatial dependence but also different ratios to each other. Asbestos fiber (crocidolite and amosite) concentration values changed in the range of 2.5 to 14.4 f per liter of air. The number of pollen grains showed that Cedar ranked higher in concentration than other types of pollen, with the maximum concentration at site A.

The results of our intensive field measurement suggested that synergic biological effects induced by ambient allergenic pollen and urban fine PM in atmosphere are associated with a peculiar springtime air quality in the Fukuoka Prefecture. Our study also indicates that the specific artificial and natural sources are regionally distributed to influence local air quality and public health. We should be mindful of the fact that in order to improve the understanding of urban air quality regarding the environmental load and repercussions on human health, it will be necessary to continue the monitoring of not only particulate matters but also gaseous materials.

Acknowledgements

The HYSPLIT aerosol dispersion model (http://www.arl.noaa.gov/ ready.html) developed by the National Oceanic Atmospheric Administration (NOAA) for backward trajectory was very helpful to data interpretation. The authors also express sincere thanks to Prof. G.-U. Kang, Department of Medical Administration, Wonkwang Health Science University for his ion analysis.

Author details

Chang-Jin Ma[1*] and Mariko Yamamoto[1,2]

*Address all correspondence to: ma@fwu.ac.jp

1 Department of Environmental Science, Fukuoka Women's University, Fukuoka, Japan

2 Fukuoka Prefectural Munakata High school, Munakata, Japan

References

[1] Takahashi Y., Tokumam K., Kawashima S. Distribution Chart of Cryptomeria Japonica Forest Through Data Analysis of Landsat-TM. Japanese Journal of Palynology 1992; 38 140-147.

[2] Schwartz J., Neas L.M. Fine Particles Are More Strongly Associated than Coarse Particles with Acute Respiratory Health Effects in Schoolchildren. Epidemiology 2000(11) 6-10.

[3] Cook P.M., Glass G.E., Tucker J.H. Asbestiform Amphibole Minerals: Detection and Measurement of High Concentrations in Municipal Water Supplies. Science 1974; 185 853-855.

[4] Camus M., Siemiatycki J., Meek M.Sc.B. Nonoccupational Exposure to Chrysotile Asbestos and the Risk of Lung Cancer. New England Journal of Medicine 1998(338) 1565-1571.

[5] Lemen R.A. Asbestos in Brakes Exposure and Risk of Disease. American Journal of Industrial Medicine 2004(45) 229-237.

[6] Monthly Archives. Fukuoka-ken in Japan Encyclopedia. http://www.asbestosremovalsaustralia.com.au/blog/2013/04/Nussbaum (accessed 2 May 2014).

[7] Fukuoka Prefecture. About Fukuoka http. http://www.pref.fukuoka.lg.jp/somu/multilingual/english/about.html (accessed 2 May 2014).

[8] Nussbaum L.F. and Käthe R. Japan encyclopedia. Cambridge: Harvard University; 2005.

[9] Deshmukh D.K., Deb M.K., Tsai Y.I., Mkoma S.L. Water Soluble Ions in $PM_{2.5}$ and PM_1 Aerosols in Durg City, Chhattisgarh, India Aerosol and Air Quality Research 2011(11) 696-708.

[10] Gao X., Nie W., Xue L., Wang T., Wang X., Gao R., Wang W., Yuan C., Gao J., Ravi K.P., Wang J., Zhang Q. Highly time-resolved Measurements of Secondary Ions in $PM_{2.5}$ During the 2008 Beijing Olympics: The Impacts of Control Measures and Regional Transport. Aerosol and Air Quality Research 2013(13) 367-376.

[11] Vezey E.L., Skvarla J.J., Vanerpool S.S. Characterizing Pollen Sculpture of Three Elosely-related Capparaceae Species Using Quantitative Image Analysis of Scanning Electron Micrographs, Clarendon: Oxford; 19991.

[12] Zhang W.J., Sun Y.L., Zhuang G.S., Xu D.Q. Characteristics and Seasonal Variations of $PM_{2.5}$, PM_{10}, and TSP Aerosol in Beijing. Biomedical and Environmental Sciences 2006(19) 461-468.

[13] Stone E.A., Yoon S.C., Schauer J.J. Chemical Characterization of Fine and Coarse Particles in Gosan, Korea During Springtime Dust events. Aerosol and Air Quality Research 2011(11) 31-43.

[14] Interim Guidance. Naturally occurring asbestos (NOA) at school sites http://www.dtsc. ca.gov/Schools/upload/SMBRP_POL_Guidance_Schools_NOA.pdf. (accessed 3 July 2014).

[15] El Dorado County. Naturally occurring asbestos and dust protection. http://www.co.el-dorado.ca.us/emd/apcd/PDF/Naturally_Occuring_Asbestos_June_12.pdf (accessed 3 July 2014).

[16] Kim J.Y., Lee S.K., Lee J.H., Lim M.H., Kang S.W., Phee Y.G. A Study on the Factors Affecting Asbestos Exposure Level from Asbestos Abatement in Building Demolition Sites. Korean Industrial Hygiene Journal 2009 (19) 8-15.

[17] Farn P.P., Clarence T.N., Patrick J.S. Aerosol characteristics of Arctic haze sampler during AGASP 2. Atmospheric Environment 1990 (24A) 937-939.

[18] Mandrioli P., Negrini M.G., Scarani C., Tampieri F., Trombetti F. Mesoscale Transport of Corylus Pollen Grains in Winter Atmosphere. Grana 1980 (19) 227-233.

[19] Saito Y., Ide T., Murayama K. Science of Pollenosis. Kagakudouzinn co; 2006. p.14-42.

8

Modelling PM2.5 with Fuzzy Exponential Membership

Danni Guo

Additional information is available at the end of the chapter

1. Introduction

Air quality is a public and environmental issue that concerns people, whether in terms of global climate change or for health and the quality of life. The Environmental Protection Agency (EPA) regulates six primary air pollutants: Ozone, Particulate matter, Carbon Monoxide, Nitrogen Oxides, Sulfur Dioxide, and Lead [3]. Particulate matter (PM) refers to solid particles and liquid droplets found in air. Many manmade and natural sources produce PM directly, or produce pollutants that react in the atmosphere to form PM [9]. $PM_{2.5}$ are small particles or particulate matter, that are less than 2.5 micrometers in diameter. $PM_{2.5}$ can be produced by combustion from motor vehicles (esp. diesel powered buses and trucks), power plants, residential wood burning, forest fires, agricultural burning, and industrial processes [15]. They can also be formed in the air through when gases (air poullutants) and orgnaic compounds are transformed through chemical reactions.

These tiny particles $PM_{2.5}$ can cause health hazards for people and also for the environment. People with heart diesease and lung problems including asthma, and also the elderly and children, are particularily vulnerable and at high risk when exposed to high levels of $PM_{2.5}$. Due to the tiny size of these $PM_{2.5}$ particles, they can penetrate to the deepest parts of the lungs, which is very dangerous to the human health. The California Air Resources Board (CARB) scentifically conducts studies and reports on the impacts of air pollutant exposures on public health, and the studies shows the negative impacts of $PM_{2.5}$ which is known to cause premature death [2]. Scientific studies has repeatedly found links between particulate matter and many health problems of people who has been exposed to high levels of $PM_{2.5}$, including asthma, bronchitis, respiratory problems, including shortness of breath and painful breathing, and premature deaths [3].

One must also be aware of the effects of high levels of $PM_{2.5}$ on the environment as well. Since $PM_{2.5}$ are tiny, they can be carried by wind and travel great distances, so that it can cause problems for areas downwind far from the actual source of air pollution. They have adverse effect on urban areas, agriculture, and the natural envrinonment. High levels of $PM_{2.5}$ can results in visibility problems, urban haze, and acid rain [3].

The U.S. Environmental Protection Agency has established standards requiring the annual average of the $PM_{2.5}$ to be not more than 15 micrograms per cubic meter [3]. The State of California monitors and reports on their air pollutants carefully, setting very high standards for their air quality ($\mu g/m^3$). From 1999-2011, there are 113 station locations monitoring $PM_{2.5}$. The site design originally planned was well spread statistically. See Figure 1. However, in reality, it is too costly in terms of time, finance, and manpower to keep all the 113 sites to be monitoring and recording every single year. Each year, only a part of the 113 sites were actually sampled, and each year at different locations.

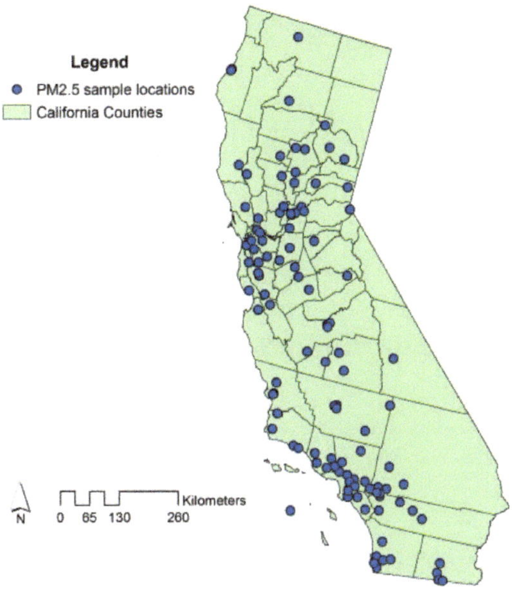

Figure 1. Complete 113 $PM_{2.5}$ Observational Sites in the California State

Comparisons of $PM_{2.5}$ between the years are difficult, due to "missing data" at sample sites [6, 9]. A site that does not have a recorded $PM_{2.5}$ value is referred to as "missing value", and since there are no patterns so that serious problems would twist the kriging map constructions.

Observing the dataset in Figure 2, the worst (in 1999) only 11 sites (9.73% of 113 sites) were used and at the best (in 2009) 65 sites (57.52% of 113 sites) were used. Over 13 years, 1469 annual arithmatic means ($\mu g/m^3$) should be recorded, but actually, 556 annual arithmatic means ($\mu g/m^3$) were reported, which occupied 37.85%. Sitewise looking, only one site, Site 2596 (Placer County APCD), was collected data annually and had 13 recored annual arithmatic means ($\mu g/m^3$), while 16 sites had one annual arithmatic mean ($\mu g/m^3$) only. The comparisons of $PM_{2.5}$ annual arithmatic means ($\mu g/m^3$) between years for a given site or between sites for a given year, i.e., the investigations of $PM_{2.5}$ annual arithmatic means ($\mu g/m^3$) patterns will be an

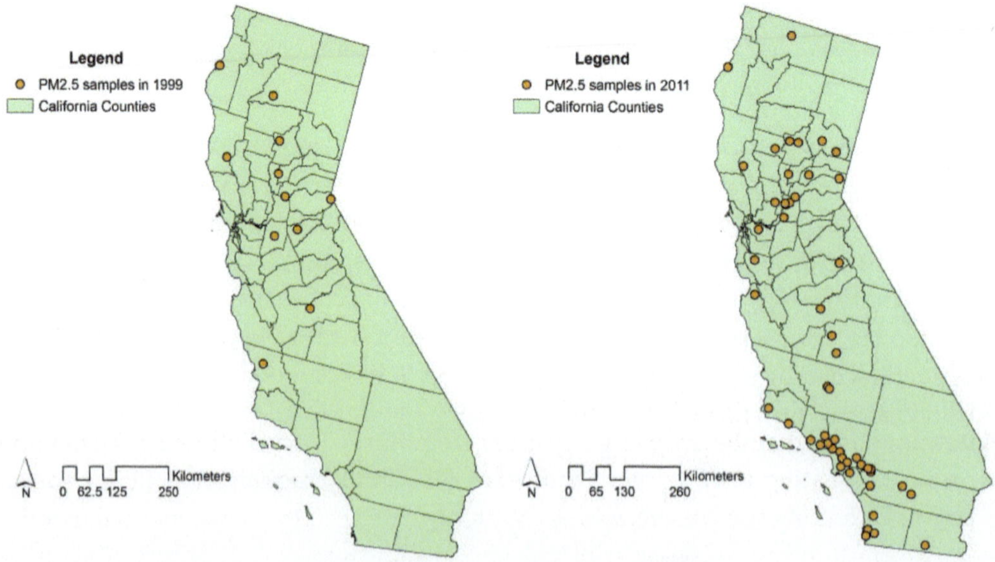

Figure 2. $PM_{2.5}$ Samples Collected in California in 1999 (11 sites) and 2011 (52 sites)

extremely difficult task due to data incompleteness. Therefore it is logical to engage fuzzy theory for treating the "missing" or scarce data.

2. Literature review and methodology

2.1. Fuzzy theory and membership kriging approach

Zadeh's fuzzy theory [22, 23] poineered a new mathematical branch. His membership approaches were quickly spreading and merging into many other mathematical branches, for example, engineering, business, economics, etc. and generating huge impacts in mathematical theories and applications. But it is aware that associated with Zadeh's fuzzy mathematical achivements, researchers gradually discovered three fundamental issues: self-duality dilemma, variable dilemma, and membership dilemma. Guo et al. [8] discussed those dilemmas in detail and pointed out Liu's credibility measure theory [11] is a solid mathematical treatment to address fuzzy phenomenon modelling. The credibility measure, similar to probability measure, assumes self-duality. Consequently, parallel to probability theory, a fuzzy variable and its (credibility) distribution can be defined. Furthermore, the membership function of a fuzzy variable can also be specified by its credibility distribution. Without any doubts, credibility measure theory is applying to practical situations sucessfully, say, Peng and Liu [14] considered parallel machine scheduling problems with processing times, Zheng and Liu [24] studied a fuzzy vehicle routing optimization problem, Guo et al. [2008] proposed credibility distribution grade kriging for investigation California State PM_{10} spatial patterns, Wang et al. [20, 21] investigated a fuzzy inventory model without backordering, Sampath and Deepa [16] developed sampling plans containing fuzziness and randomness, and others. Nevertheless, the investigations in statistical estimation and hypothesis testing problems with fuzzy credibility distribution are very slowly progressed, for example, Li et al. [10], Sampath and Ramya [17, 18] in worked with fuzzy normal distributions, and studied the exponential credibility distribution function [1].

With statistically well-designed scheme, the collected spatial data can be analyzed and presented by kriging maps. If we are facing spatial data with "missing" or scarce fuzziness, it is impossible to construct kriging maps. It is noticed that the air pollutants were measured in different sites each year, even the site design originally planned was well spread statistically. We call a site that does not have a recorded $PM_{2.5}$ concentration as "missing value" site, as continued from Guo's research [4, 5]. To address the basic requirements in constructing kriging maps, Guo first proposed membership kriging in Zadeh's sense, see the MSc thesis [4] in which the linear, quadratic and hyperbolic tangent membership functions were used. Later Guo [5, 7] had developed the membership kriging under credibility theory. Following the membership kriging route of treating uncertainty, Shada et al. [19] and Zoraghein et al. [25] made considerable contributions in their papers. In this chapter, we will integrate exponential membership and kriging, to fill in the "missing data" based on existing sample data, and making a comparison of $PM_{2.5}$ concentrations in California from 1999-2011.

2.2. Fuzzy exponential distribution

It is necessary to introduce the basics of Liu's fuzzy credibility theory [11]. Let Θ be a nonempty set and M a σ- algebra over Θ. Elements of M are called events. $\mathrm{Cr}\{A\}$ denotes a number or grade associated with event A, called credibility (measure or grade). Credibility measure $\mathrm{Cr}\{\cdot\}$ satisfies the axioms normality, monotonicity, self-duality and maximality:

$$\begin{aligned}
&\mathrm{Cr}\{\Theta\}=1\\
&\mathrm{Cr}\{A\}\le\mathrm{Cr}\{B\}\ \text{ for } A\subset B\\
&\mathrm{Cr}\{A\}+\mathrm{Cr}\{A^c\}=1 \text{ for any event } A\\
&\mathrm{Cr}\left\{\bigcup_i A_i\right\}=\sup_i\left(\mathrm{Cr}\{A_i\}\right) \text{ for any events } A_i \text{ with } \sup_i\left(\mathrm{Cr}\{A_i\}\right)<0.5
\end{aligned} \tag{1}$$

Definition 1: The set function Cr{} on M is called a credibility measure if it follows Axiom Normality, Axiom Monotonicity, Axiom Self-Duality and Axiom Maximality shown in Equation 1. The triplet (Θ, M ,Cr) is called a credibility space.

Definition 2: A measurable function,η, mapping from a credibility space (Θ, M,Cr) to a set of real numbers ῷ 𝔹? $\subset \mathbb{R}$.

Definition 3: The credibility distribution Ψ of a fuzzy variable η on the credibility measure space (Θ, M,Cr) is Ψ: $\mathbb{R} \rightarrow$ [0,1], where $\Psi(x) = \mathrm{Cr}\{\theta\in\Theta \mid \eta(\theta)\ x\}$, x $\in \mathbb{R}$.

Definition 4: The function μ of a fuzzy variable η on the credibility measure space (Θ, M,Cr) is called as a membership function: $\mu(x)=(2\mathrm{Cr}\{\eta=x\})\wedge 1,\ \ x\in\mathbb{R}$.

Theorem 5: (Credibility Inversion Theorem) Let η be a fuzzy variable on the credibility measure space (Θ, M,Cr) with membership function μ. Then for any set B of real numbers,

$$\mathrm{Cr}\{\eta \in B\} = \frac{1}{2}\left(\sup_{x\in B}\mu(x) + 1 - \sup_{x\notin B}\mu(x)\right). \tag{2}$$

Corollary 6: Let η be a fuzzy variable on a credibility measure space (Θ, M,Cr) with membership function μ. Then the credibility distribution Ψ is

$$\Psi(x) = \frac{1}{2}\left(\sup_{y\le x}\mu(y) + 1 - \sup_{y>x}\mu(y)\right), \text{ for } \forall x \in ¡ . \tag{3}$$

It is obvious the concept of credibility measure is very abstract although the credibility measure has normality, monotonicity, self-duality and maximality mathematical properties. The credibility measure loses the intuitive feature as Zadeh's membership orientation. Credibility Inversion Theorem and its corollary have just revealed the deep link between an abstract measure and intuitive membership. Such a link definitely paves the way for real-life applications. For example, the trapezoidal fuzzy variable has membership function μ:

$$\mu(x) = \begin{cases} 0 & x \le c - c_2 \\ \dfrac{c_2 + x - c}{c_2 - c_1} & c - c_2 < x \le c - c_1 \\ 1 & c - c_1 < x \le c + c_1 \\ \dfrac{c_2 - x + c}{c_2 - c_1} & c - c_1 < x \le c + c_1 \\ 0 & c + c_2 < x \end{cases} \tag{4}$$

where c, c_1, c_2 are all positive200, $c > c_1$,$c > c_2$, $c_2 > c_1$.

With the help of Equation 3, the fuzzy credibility distribution is thus,

$$\Psi(x) = \begin{cases} 0 & x \le c - c_2 \\ \dfrac{c_2 + x - c}{2(c_2 - c_1)} & c - c_2 < x \le c - c_1 \\ \dfrac{1}{2} & c - c_1 < x \le c + c_1 \\ \dfrac{2c_2 - c_1 + x - c}{2(c_2 - c_1)} & c - c_1 < x \le c + c_1 \\ 1 & c + c_2 < x \end{cases} . \tag{5}$$

Liu [12, 13], Wang and Tian [21] and [1] studied the exponential fuzzy distribution with a membership function, denoted as Exp(m):

$$\mu(x)=\frac{2}{1+e^{\frac{\pi x}{\sqrt{6}m}}},x\geq 0,m>0 \tag{6}$$

The support of an exponential membership function is set [0,+∞), the nonnegative part of the real line, ℝ. The expected value and second moment of exponential fuzzy variable are

$$\mathrm{E}[\eta]=\frac{\sqrt{6}\ln 2}{\pi}m \tag{7}$$

and

$$\mathrm{E}\left[\eta^2\right]=m^2 \tag{8}$$

where the parameter $m > 0$ determining the mean and variance of the exponential fuzzy variable. Thus it is intuitive to reveal how the shape of membership curve affected by the possible values of the parameter $m > 0$.

x	m=12	m=15	m=22	m=50
0	1.0000	1.0000	1.0000	1.0000
1	0.9466	0.9573	0.9709	0.9872
2	0.8935	0.9147	0.9418	0.9744
3	0.8410	0.8724	0.9128	0.9615
4	0.7894	0.8306	0.8839	0.9487
5	0.7390	0.7894	0.8553	0.9360
6	0.6899	0.7490	0.8269	0.9232
7	0.6424	0.7094	0.7987	0.9105
8	0.5968	0.6707	0.7709	0.8978
9	0.5530	0.6332	0.7435	0.8851
10	0.5113	0.5968	0.7165	0.8724
11	0.4717	0.5616	0.6899	0.8598
12	0.4342	0.5277	0.6638	0.8473
13	0.3990	0.4952	0.6382	0.8348
14	0.3660	0.4640	0.6132	0.8223
15	0.3351	0.4342	0.5887	0.8100
16	0.3063	0.4059	0.5647	0.7976
17	0.2796	0.3789	0.5414	0.7854

x	m=12	m=15	m=22	m=50
18	0.2549	0.3533	0.5187	0.7732
19	0.2320	0.3292	0.4966	0.7610
20	0.2110	0.3063	0.4752	0.7490
21	0.1917	0.2848	0.4544	0.7370
22	0.1739	0.2645	0.4342	0.7251
23	0.1577	0.2455	0.4147	0.7133
24	0.1428	0.2277	0.3959	0.7016
25	0.1293	0.2110	0.3777	0.6899
26	0.1170	0.1954	0.3602	0.6784
27	0.1057	0.1808	0.3433	0.6669
28	0.0955	0.1672	0.3270	0.6556
29	0.0863	0.1546	0.3114	0.6443
30	0.0779	0.1428	0.2964	0.6332

Table 1. Impacts of m in the Shape of Exponential Membership Curve

The value choice of parameter m is not aimless. m = 22 corresponds to the $PM_{2.5}$ annual arithematic mean 11.85 (μg/m^3), while m = 50 corresponds to the $PM_{2.5}$ annual arithematic mean 25.89 (μg/m^3). Therefore, the one-parameter exponential fuzzy variable may well cope to the modelling requirements of the California $PM_{2.5}$ annual arithematic mean dataset. Furthermore, using the one-parameter exponential fuzzy variable it may develop a delicate scheme of testing credibility hypothesis.

2.3. Credibility hypothesis testing with exponential membership

Similiar to the popular Neyman-Pearson Lemma in probability theory, likelihood ratio L_0/L_1, constant k, and critical region C of size α, are involved in the testing hypothesis H_0: $\theta = \theta_0$ against alternative hypothesis H_1: $\theta = \theta_1$. The likelihood is defined as the product of the densities for given sampled population. Hence we can call Neyman-Pearson testing criterion as likelihood ratio criterion. Inevitably, Type I error and Type II error concepts are also engaged in describing testing procedure. For hypothesis testing under credibility theory, Sampath and Ramya [17] proposed a membership ratio criterion. The membership criterion applies to any forms of credibility distributions, but exponential credibility distribution has its unique advantage. Therefore, the remaining descriptions will be focused on credibility hypothesis testing under exponential membership function [1].

Definition 7: Credibility hypothesis is a statement describing the possible rejection of a null hypothesis, H_0: $\mu = \mu_0$ with the credibility distribution of a fuzzy variable against an alternative hypothesis H_1: $\mu = \mu_1$ with another credibility distribution of a fuzzy variable.

Definition 8: Credibility hypothesis testing is the rule describing reject or not reject a null hypothesis if the calculated values sampled from the fuzzy distribution defined by null hypothesis.

Definition 9: Credibility rejection region is the subset of the support under a fuzzy distribution, denoted as C, on which the null credibility hypothesis is rejected H_0: $\mu = \mu_0$, i.e., $C = \{\eta \in \Theta \mid H_0 \text{ is rejected}\}$

Definition 10: Type I error is the mistake by rejecting the null credibility hypothesis H_0: $\mu = \mu_0$ when it is true and Type II error the mistake by not rejecting the null credibility hypothesis H_0: $\mu = \mu_0$ when it is false.

Definition 11: Level of credibility significance is the maximal credibility to make a Type I error in testing a credibility hypothesis H_0: $\mu = \mu_0$, denoted as α.

Definition 12: The best credibility rejection region of credibility significance level α, C^*, if this region possesses the maximal power (measured by credibility) under alternative hypothesis K with all possible credibility rejection regions of level of credibility significance α, i.e.,

$$\mathrm{Cr}\{\eta \in C^* \mid K\} \geq \mathrm{Cr}\{\eta \in C \mid K\}, \tag{9}$$

where C is any region satisfying the condition $\mathrm{Cr}\{\eta \in C \mid H\} \leq \alpha$. The power of the credibility hypothesis testing is $\mathrm{Cr}\{\eta \in C^*\}$.

With the exponential membership function having single parameter $m > 0$, the best credibility rejection region of credibility significance level α, C^* should be an interval so that we name it as best credibility rejection interval of credibility significance level α.

Theorem 13: For testing the null credibility hypothesis H_0: $m = m_1$ against the alternative credibility hypothesis K: $m = m_2$, $(m_1 < m_2)$, under exponential fuzzy distributions $\mu(x) = \mathrm{Exp}(m)$ as Equation 6 specified, the membership ratio criterion is engaged. The criterion states that given credibility significance level $\alpha < 0.5$, the best credibility rejection interval C^*:

$$C^* = \left[\frac{\sqrt{6}m_1}{\pi}\ln\left(\frac{1-\alpha}{\alpha}\right), +\infty\right), \ \alpha \in (0, 0.5). \tag{10}$$

The credibility of credibility rejection interval C^* under alternative hypothesis is greater than α.

α	0.10	0.15	0.20	0.25	0.30
x_0	37.56361	29.65463	18.78181	14.48535	10.58305

Table 2. x_0 and α under H_0: m_1=21.926

Table 2 illustrates relationship between the best credibility rejection interval boundary x_0 for selected credibility significance level $\alpha < 0.5$. For example, let $\alpha = 0.20$, $m_1 = 21.93$, then the best credibility rejection interval C^*= [18.78, +∞). We have to point out that the choice of credibility significance level α in credibility hypothesis testing should not follow the "thumb rule" the significance level α = 0.05 in probability hypothesis testing. Although the two significance levels have the same role in hypothesis testing, nevertheless, the practical meanings of credibility significance level α and the significance level α are quite different. From Table 2 and California $PM_{2.5}$ distribution patterns, it is is logical and practical selecting the credibility significance level α = 0.25, which gives x_0 =14.485.

3. Analysis and results

3.1. Exponential fuzzy membership kriging

Now having examined the methodology, we can now calculate the membership grades. But first let us have a quick overlook on California $PM_{2.5}$ 1999-2011 records.

Year	Number of Sites	Annual Maximum	Annual Minimum	Annual Average
1999	11	23.4190	8.3274	13.6146
2000	22	23.9229	7.7971	14.3465
2001	25	25.0446	4.1492	12.5643
2002	41	25.8332	6.3356	15.9160
2003	44	24.8348	4.3802	12.6993
2004	43	18.1962	4.4146	11.3249
2005	40	22.4178	4.8117	12.0750
2006	47	21.5991	5.1022	11.4211
2007	55	25.2073	3.2883	11.6240
2008	59	21.2357	6.7716	12.3861
2009	65	21.2423	3.3373	10.7292
2010	52	17.1908	2.9665	9.2255
2011	52	18.0950	3.3682	10.7934
Total	556			

Table 3. California $PM_{2.5}$ 1999-2011 Annual $PM_{2.5}$ Concentrations

From Table 3, we can see that the annual average $PM_{2.5}$ range between 9.2255 and 15.9160. Most of the averages are wondering about 11.0 and 12.7. Therefore, it is very reasonable to estimate $\overline{X}$ =11.85.

Exponential membership grade kriging scheme:

1. Calculating overall exponential membership parameter m. Based on Equation 7, we can estimate the exponential membership parameter m. Then in this paper, m = 21.93 is used for membership grade smoothing, kriging, and hypothesis testing.

2. Calculating every exponential membership grade $e_{ij} = 2.0/(1+\exp(\pi x_{ij}/m\sqrt{6}))$ for site j given year i. For any site, if "missing" value, $x_{ij} = 1$, otherwise, $x_{ij} = e_{ij}$ (> 0).

3. Performing membership grade interpolation. (i) Equal weights, for any given site j, if x_{ij} = 1, which is a "missing" value, let $x_{ij} \triangleq \hat{e}_{ij}$ where $\hat{e}_{ij} = (e_{i-1,j} + e_{i+1,j})/2$ (conditioning on the availability of the nearest neighbours $e_{i-1,j}$ and $e_{i+1,j} < 1$); (ii) Unequal weights, for a given site j, if the neighbour years are quite far, say, N between the gap, then, we may take linear interpolation for filling those "missing" value. that is given e_{ij} and $e_{i+N,j}$ < 1, let $x_{i+l,j} \triangleq e_{ij} + l * (e_{i+N,j} - e_{ij})/N$, $l = 1,2,\ldots, N\text{-}1$.

4. Performing membership grade extrapolation, including forward and backward extrapolation. Equal weights (1/3) are used mostly. Unequal weights are also used.

5. Carrying the filling "missing" cells task until thirteen years 1999 to 2011, each year 113 membership grades are all calculated, $\{\hat{e}_{ij}, i=1, 2, \cdots, 13, j=1, 2, \cdots, 113\}$.

Now, it is ready to construct exponential membership grades kriging maps and use these 13 maps for comparisons.

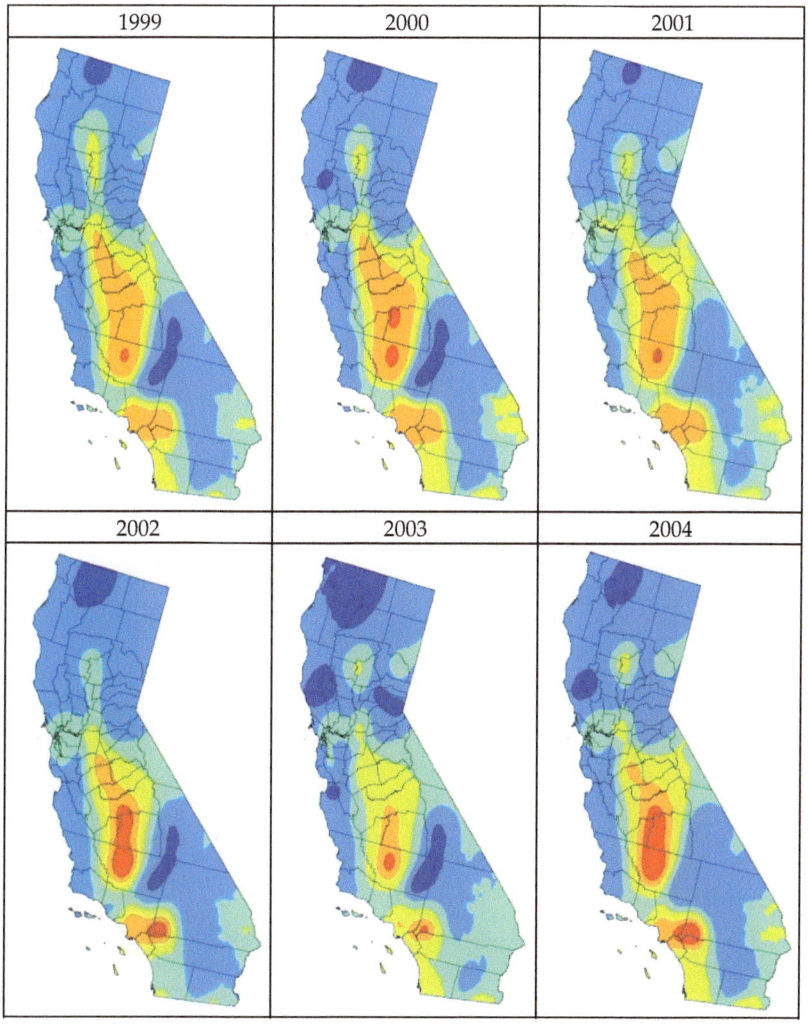

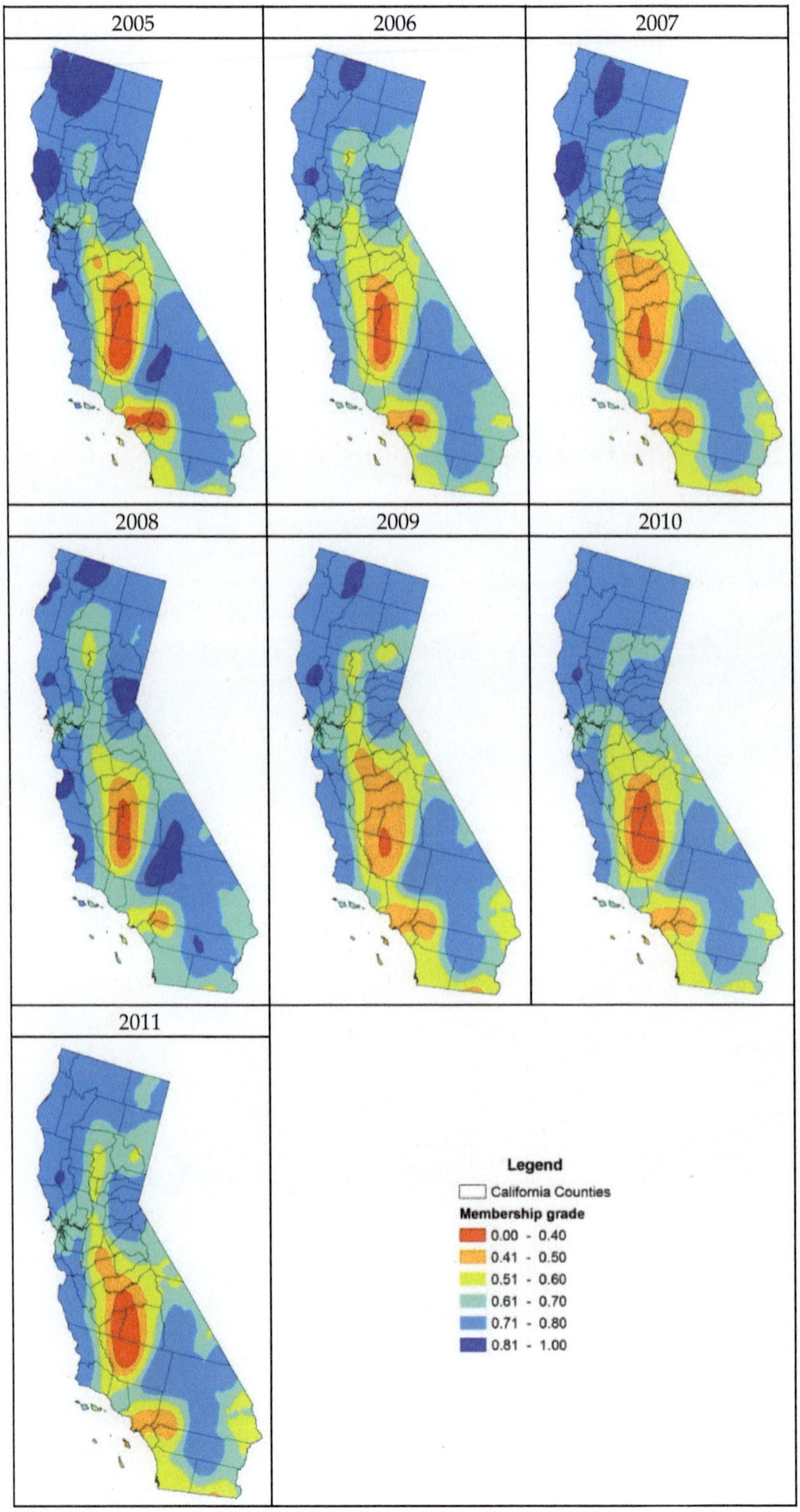

Figure 3. Kriging maps of Exponential membership grades for California 1999-2011

As one can observe from Figure 3, central California regions have very low membership grades, and the rest of Calfornia have higher membership grades.

3.2. Calculated $PM_{2.5}$ concentrations

It is impossible for the public and governmental officers to accept the membership grade kriging maps. Therefore, kriging maps of every year $PM_{2.5}$ concentrations (collected and estimated together) must be constructed. The conversion formula is

$$\tilde{x}_{ij} = \frac{\sqrt{6}m}{\pi}\ln\left(\frac{2-\hat{e}_{ij}}{\hat{e}_{ij}}\right). \tag{11}$$

Kriging maps with "completed" of every year $PM_{2.5}$ data $\{x_{ij}, i=1,2,...,13, j=1,2,...,113\}$. Comparison of kriging maps with the "completed" data can now be performed. See Figure 4. A complete dataset is now available, the 113 observation sites now have the full 13 year $PM_{2.5}$ concentration, containing 1469 data records. The 13 ordinary kriging prediction maps are generated, and one can clearly observe the changes in $PM_{2.5}$ year by year.

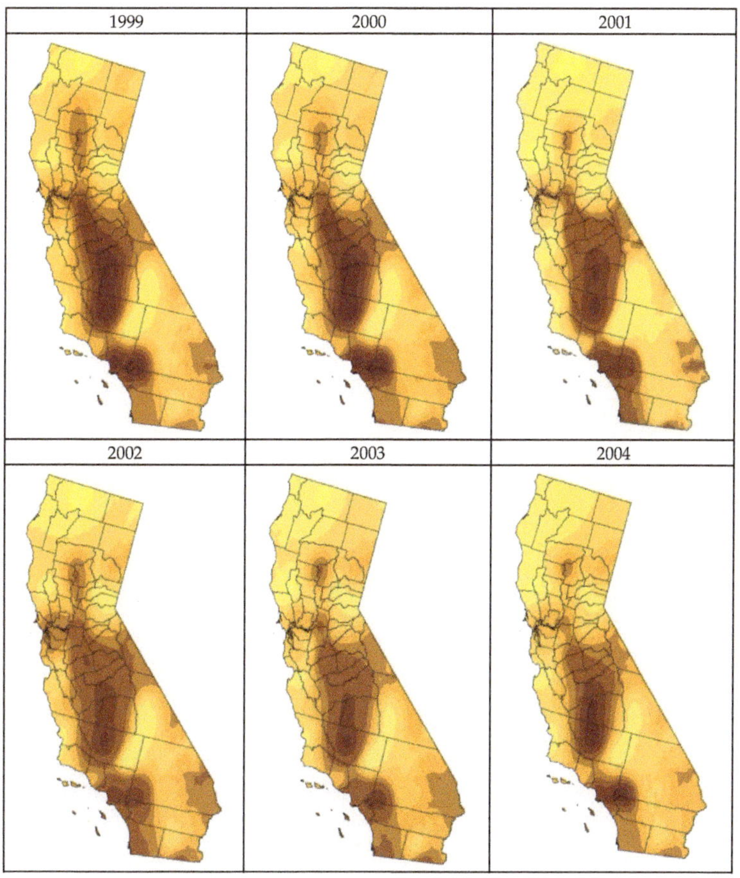

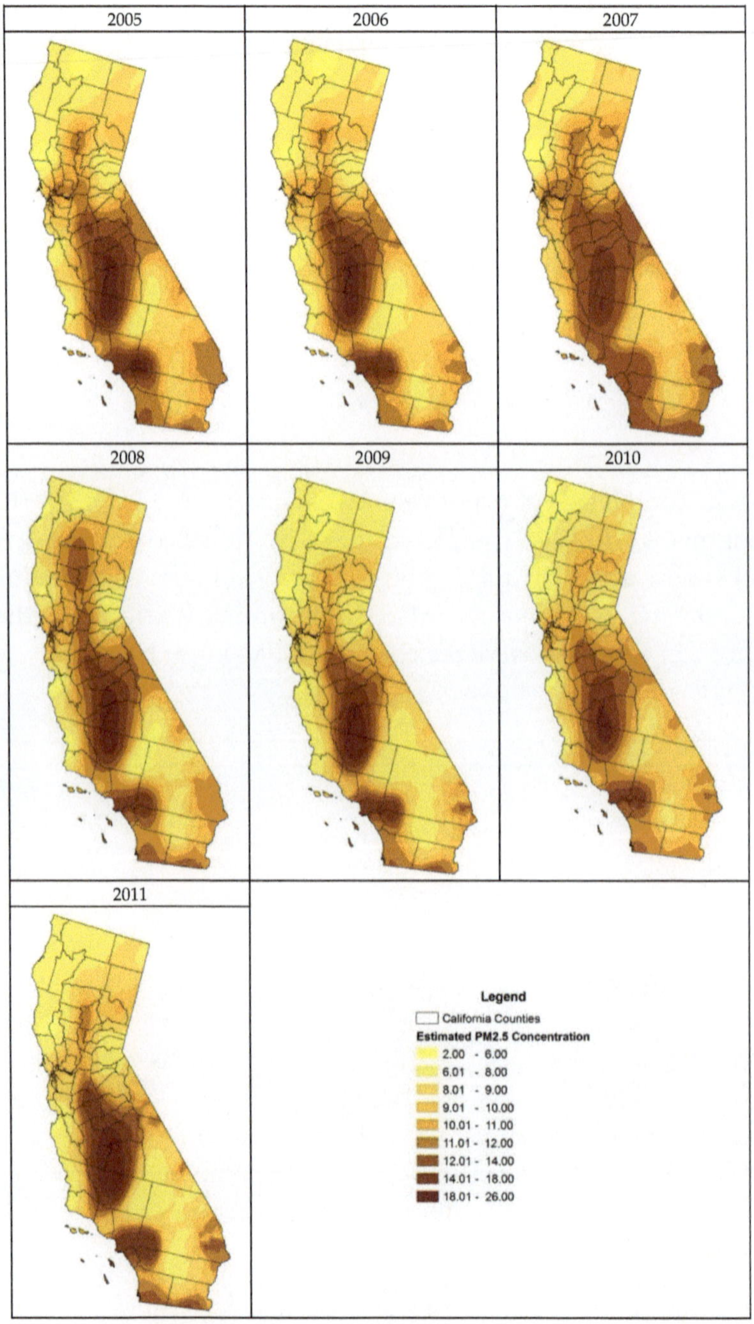

Figure 4. Kriging maps of completed estimated $PM_{2.5}$ concentrations for California 1999-2011

One can now observe and compare the year by year changes of $PM_{2.5}$ concentration, and note the regions of high and low $PM_{2.5}$ concentration. The dark brown colours represent high $PM_{2.5}$ concentrations, and light yellow colours represent areas with low $PM_{2.5}$ concentrations. Note that central California shows to have continual high levels of $PM_{2.5}$ concentration, year after year.

4. Interpretation and conclusion

In the results section, the dataset is now calculated and completed. However, it is now up to us to interpret the maps, and decide how to best make use of the calculated dataset, so that it provides us with easy to read information. And we can do this through a change map and 13 health safety maps.

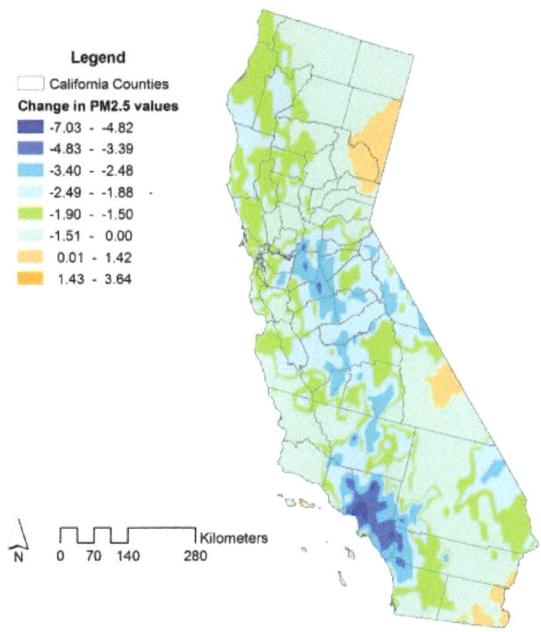

Figure 5. Changes in $PM_{2.5}$ Concentrations in California between 1999 and 2011

As one can clearly see from Figure 5, that $PM_{2.5}$ concentration has clearly decreased and air quality has improved remarkably over the years. The blue and green colours show negative changes, and orange shows positive changes. Counties such as Los Angeles and Orange show the highest decrease, and other counties such as Lassen, Plumas, Sierra, Inyo, and Imperial show some increase in $PM_{2.5}$ concentration. However, *a decrease in* $PM_{2.5}$ *concentration does not indicate safety in air quality.*

In terms of credibility hypothesis testing, say, with credibility significance level $\alpha = 0.25$, critical point for the best credibility rejection interval is $x_0 = 14.485$. The indictor λ_{ij} is defined as

$$\lambda_{ij} = \begin{cases} 1 & \tilde{x}_{ij} < 14.485 \\ 0 & \text{othewise} \end{cases}, \tag{12}$$

For comparisons of air quality safety, we generate $PM_{2.5}$ safety maps with two colours: blue colour if $\lambda_{ij} = 1$, orange colour otherwise, in total 13 safety maps. See Figure 6.

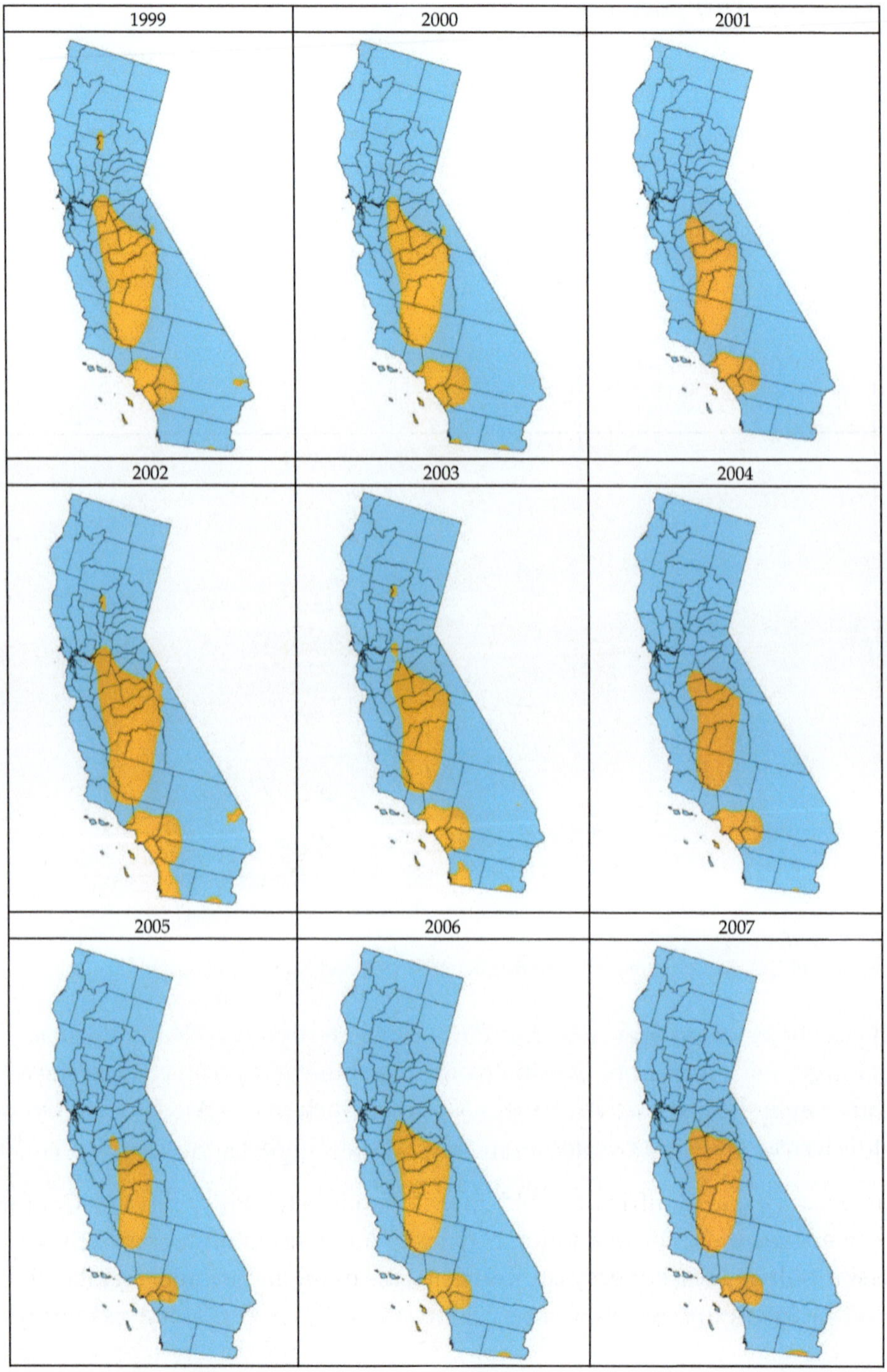
1999
2000
2001
2002
2003
2004
2005
2006
2007

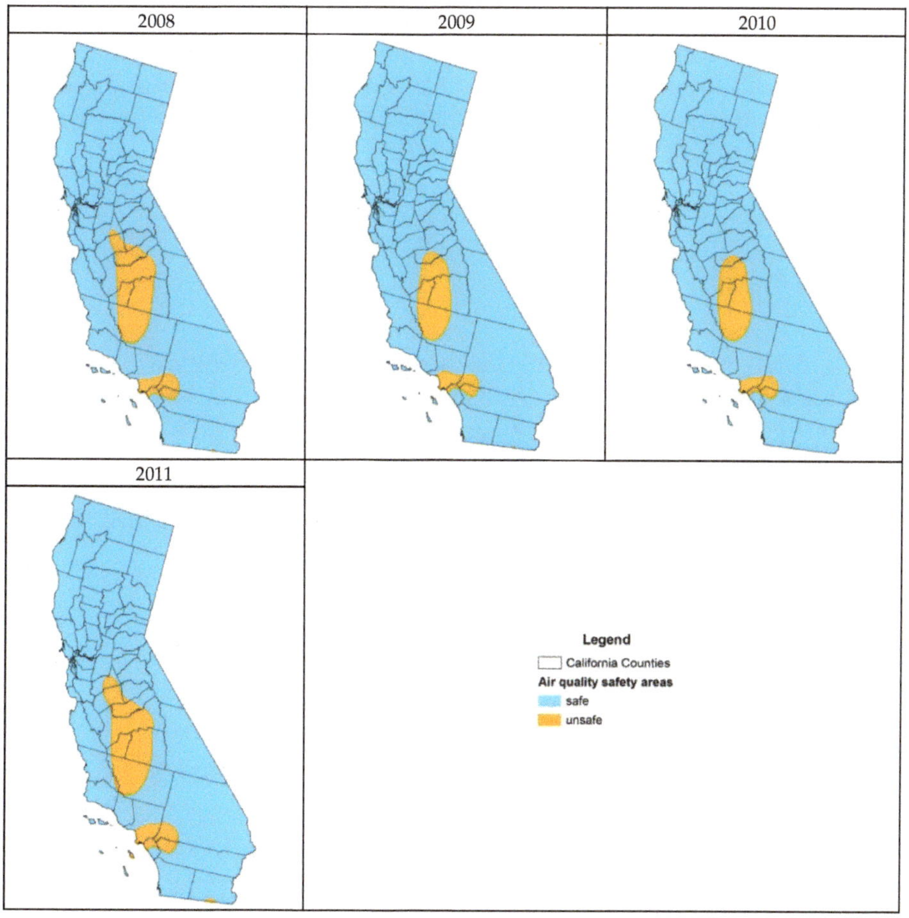

Figure 6. Air Quality $PM_{2.5}$ Safety areas in California 1999-2011

One can now observe that over the 13 years period 1999-2011, Stanislaus, Merced, Madera, Fresno, Kings, Tulare, Kern, Los Angeles, San Bernardino, Orange, and Riverside counties are the counties with the highest $PM_{2.5}$ safety problems. These areas have shown to be unsafe for public health safety, and especially for those with lung and heart problems, and for children and the elderly. These places are also sources of environmental and ecological concerns.

In conclusion, facing the difficult problem of "incomplete" $PM_{2.5}$ data in California from 1999-2011, we used the interpolation and extrapolation smoothing approaches for "filling" those "missing value" sites. For easy computation, the fuzzy exponential membership function is assumed. The treatment is based on an assumption that the smoothing is performed for a given site rather than over different sites for a given year. Such an assumption is emphasizing the fact: the data recorded are $PM_{2.5}$ concentration annual arithmetic means and they shouldn't change too dramatically over neighbour years. As to neighbour sites impacting, the membership grade kriging approach is adequate enough for generating smoothed maps. Furthermore, for utilizing credibility hypothesis testing theory, we perform parameter estimation of the fuzzy exponential membership function and in terms of membership ratio criterion for deriving the safety maps under 0.25 credibility significance level. Membership ratio criterion is very similar to likelihood ratio criterion in theoretical development. By comparing those 13 $PM_{2.5}$ concentration safety maps to 1999-2011 change map, it is quite justifiable to say the safety maps under the credibility hypothesis testing procedure are very intuitive and convenient to the public. Finally, interpreting the 13 safety maps will provide the public with knowledge of air quality in California.

Acknowledgements

I would like to thank the California Air Resources Board for providing the air quality data used in this paper. This study is supported financially by the National Research Foundation of South Africa (Ref. No. IFR2009090800013).

Author details

Danni Guo*

Address all correspondence to: d.guo@sanbi.org.za

Statistical Ecology Division, Biodiversity Research, Assessment and Monitoring, South African National Biodiversity Institute, Cape Town, South Africa

References

[1] Authors unknown. Credibility Hypothesis Testing of Fuzzy Exponential Distributions. International Journal of Fuzzy Computation and Modelling (IJFCM). (*accepted*)

[2] California Environmental Protection Agency Air Resources Board (CARB). Ambient Air Quality Standards (AAQS) for Particulate Matter. www.arb.ca.gov (accessed 3 July 2014)

[3] Environmental Protection Agency (EPA). National Ambient Air Quality Standards (NAAQS). U.S. Environmental Protection Agency. www.epa.gov (accessed 3 July 2014)

[4] Guo D. Integrating GIS with Fuzzy Logic and Geostatistics: Predicting Air Pollutant PM10 for California, Using Fuzzy Kriging. MSc thesis. University of Durham; 2003.

[5] Guo D. Contributions to Spatial Uncertainty Modelling in GIS: Small Sample Data. PhD Thesis. University of Cape Town; 2007.

[6] Guo D., Guo R., Cui YH., Midgley GF., Altwegg, R., Thiart, C. Climate Change Impact on Quiver Trees in Arid Namibia and South Africa. Blanco, Kheradmand (ed.)

Climate Change - Geophysical Foundations and Ecological Effects. InTech; 2011. p323-342.

[7] Guo D, Guo R, Thiart C. Predicting Air Pollution Using Fuzzy Membership Grade Kriging. Computers, Environment and Urban Systems. 2007 31(1) 33-51.

[8] Guo D, Guo R, Thiart C. Credibility Measure Based Membership Grade Kriging. International Journal of Uncertainty, Fuzziness and Knowledge-Based Systems. 2007 15(2) 53-66.

[9] Guo D., Guo R., Thiart C., Cui YH. Imprecise Uncertainty Modelling of Air Pollutant PM10. In: Nejadkoorki (ed.) Advanced Air Pollution. InTech; 2011. p193-212.

[10] Li X., Qin Z, Ralescu D. Credibilistic parameter estimation and its application in fuzzy portfolio selection. Iranian Journal of fuzzy systems. 2011 15(2) 57-65.

[11] Liu B. Uncertainty Theory: An Introduction to Its Axiomatic Foundations. Springer-Verlag Heidelberg, Berlin. 2004.

[12] Liu B. Uncertainty Theory. 3rd ed. Springer-Verlag Berlin. UTLAB; 2010.http://orsc.edu.cn/liu/ut.pdf

[13] Liu B. Uncertainty Theory. 4th ed. Springer-Verlag Berlin. UTLAB; 2012. http://orsc.edu.cn/liu/ut.pdf

[14] Peng J, Liu B. Parallel machine scheduling models with fuzzy processing times. Information Sciences. 2004 166(1-4) 49-66.

[15] Pittsburghtoday. Environment / PM2.5. Pittsburghtoday.org. www.pittsburghtoday.org/CurrentPM25US (accessed 3 July 2014)

[16] Sampath S, Deepa SP. Determination of Optimal Chance Double Sampling Plan using Genetic Algorithm. Model Assisted Statistics and Applications. IOS Press. 2013 8(4) 265-273.

[17] Sampath S, Ramya B. Credibility hypothesis testing of expectation of fuzzy normal distribution. Journal of Intelligent & Fuzzy Systems. 2014.

[18] Sampath S, Ramya B. Credibility hypothesis testing of fuzzy triangular distributions. Journal of Uncertain Systems. (*accepted*)

[19] Shada R, Mesgarib MS, Abkarc A, Shad A. Predicting air pollution using fuzzy genetic linear membership kriging in GIS. Computers, Environment and Urban Systems. 2009 33(6) 472-481.

[20] Wang X, Tang W, Zhao R. Fuzzy Economic Order Quantity Inventory Models without Backordering. Tsinghua Science and Technology. 2007 12(1) 91-96.

[21] Wang Z, Tian F. A Note of the Expected Value and Variance of Fuzzy Variables. International Journal of Nonlinear Science. 2010 9(4) 486-492.

[22] Zadeh LA. Fuzzy sets. Information and Control. 1965 8 338-353.

[23] Zadeh LA. Fuzzy sets as a basis for a theory of possibility. Fuzzy Sets and Systems. 1978 1 3-28.

[24] Zheng Y. Liu B. Fuzzy vehicle routing model with credibility measure and its hybrid intelligent algorithm. Applied Mathematics and Computation. 2006 176(2) 673-683.

[25] Zoraghein H, Alesheikh A A, Alimohammadi A, Vahidnia, M H. The utilization of soft transformation and genetic algorithm to model two sources of uncertainty of Indicator Kriging. Computers, Environment and Urban Systems 2012 36(6) 592-598.

9

Air Pollution in High-Risk Sites–Risk Analysis and Health Impact

M. Ferrante, M. Fiore, C. Copat, S. Morina, C. Ledda, C. Mauceri and G. Oliveri Conti

Additional information is available at the end of the chapter

1. Introduction

During the last century, at the top of industrialization phase, the excessive use of increasingly more complex inorganic and organic chemicals has caused serious environmental contamination all over the world with acute and chronic effects on population health.

The sites considered to be very contaminated and classified as dangerous are very numerous and often belong to industrial centres; in the past years, the contamination of soil and their contribution to air pollution have had an important role.

In countries characterized by a high socio-economic status, the industrial impact on air pollution is certainly less than the weight of traffic and urban heating, while in the developing and emerging countries, industrialization has still an important and unresolved role because of the lack of legislative management and control.

Epidemiological studies represent the scientific basis used to verify the existence of negative health effects caused by air pollution and to quantify the value of these effects, estimating the dose-response relationships. Many studies have demonstrated that air pollution is associated with a wide range of adverse health outcomes. Statistical analyses, conducted on monitoring and biomonitoring data, also show a relationship between the level of air pollution and the rate of mortality and morbidity. Unlike accidents, pollution is not considered a cause of immediate death. Actually, it is thought that it represents a cause of premature death. Air pollution is associated with a great number of illnesses, such as cancer, respiratory, cardiovascular and neurological diseases, and also diabetes and infertility problems.

The estimated cost of pollution becomes an economic evaluation of the risk of getting sick or dying prematurely. But the effects of pollution are particularly difficult to evaluate in terms of quantity and cost. In order to assess the related risk of air pollution, it is certainly very important to establish the effects of pollution on human health, but it is also needful to consider some health determinants such as genetic factors, age, sex, and lifestyle habits.

Considering all these aspects, our proposal includes a chapter about atmospheric and urban pollution describing their relative and assessed effects on human health, with a detailed regard on risk assessment. In the last decades, a large number of legislations have been issued accepting international subscribed agreement, such as the Kyoto protocol, to save public health and restrict the effects of pollution on climate change. We will also focus on indoor pollution, with its insufficient data and relative problems and difficulties in management, on methodologies of study, and on some case reports.

2. Description of air pollution and characterization of monitored atmospheric pollutants

Air pollution, both for indoor or outdoor environments, is caused by several causes like chemical, physical, or biological modification of the natural characteristics of the atmosphere. Most common household combustion devices, motor vehicles, and industrial process are anthropogenic sources of air pollution, but also natural sources are very important for air pollution assessment as the forest fires and volcanoes emissions. Particulate matter, carbon monoxide, ozone, nitrogen dioxide, and sulphur dioxide today are the most important pollutants for their serious health risks [1].

In Europe, emissions of many air pollutants have decreased significantly the last 30 years, and air quality has improved across the region. However, excess of air quality standards still occurs, especially in metropolitan areas, and air pollutants released in one country may be transported in the atmosphere, polluting or degrading the air quality in neighbouring countries. On the basis of this statement, air pollution can be considered as a local, pan-European, and hemispheric issue [2].

At present, particulate matter, nitrogen dioxide, and ground-level ozone are recognized as the three most dangerous pollutants that affect human health, ranging in severity of impact according to long-term and peak exposures. Moreover, benzo(a)pyrene is a carcinogenic substance of increasing concern, as its concentrations are above the threshold set to protect human health in several urban areas, especially in central and eastern Europe. Air pollution could also represent a danger for the environment, producing acidification, eutrophication, and damage to agriculture.

There are various sources of air pollution, both anthropogenic and of natural origin, in turn divided in stationary and mobile sources. The most common anthropogenic sources of pollutants are combustion processes used for electricity generation, transport, industry and households, industrial processes and solvent use, agricultural crops and livestock, waste incinerator and landfill. A stationary source of air pollution, also known as a point source, refers to a fixed emission source, represented by factories, power plants, dry cleaners, residential wood burners, dry lakebeds, and landfills. A mobile source of air pollution is a source capable of moving under its own power. "On-road" transportation is related to all vehicles; however, there is also a "non-road" or "off-road" transportation, including the gas-powered machines for garden maintenance or recreational vehicles.

People or their activities do not cause natural "air pollution," and emission sources generally comprise volcanic eruptions, windblown dust, sea-salt spray, wild animals in their natural habitat, and plants releasing volatile organic compounds [2].

In the evaluation of pollution effects, it must be kept in mind that dangerous substances can be emitted as primary or secondary pollutants. Primary pollutants are emitted directly into the air from pollution sources, while secondary pollutants are formed when primary pollutants undergo chemical changes in the atmosphere. For example, NO_X and SO_2 are directly emitted into the air following fuel combustion or industrial processes. In contrast, O3 and the major part of PM form in the atmosphere following emissions of various precursor species, and their concentrations depend strongly on meteorological conditions, such as high air temperatures and sunlight.

At present, the attention of scientist and legislator is focused on the major sources of dangerous substances, represented by urban and industrial activities.

In the evaluation of urban pollution, it is important to consider regional pollution, city pollution, real and *hot spot* events, determined by peaks of pollutants higher than average followed by gradual restoration of normal limits. Of these three components, the first two have direct effects on human and environmental health, as they expose population to contaminants for a longer time, causing chronic effects.

As regards industrial pollution, there are various unwanted substances and losses generated by industrial activities both for qualitative and quantitative aspects [3].

For this reason, to assess significant trends and to discern the effects of reduced anthropogenic precursor emissions, long time series of measurements and a continuous monitoring of air quality are needed [4].

The application of European Directives contained the emission of many pollutants, consequently reducing their levels in the atmosphere. However, while restrictions are imposed on industries, technologies are developed to reduce pollutants from road transport and the most harmful substances in the air are continuously monitored, air pollution is still a major problem and it is necessary to focus attention on its effects on human health and on methods to reduce these effects.

At present, based on *"The Clean Air Act"*, Quality Standards are set for six common air pollutants, also known as "criteria pollutants". They are particulate matter, ground-level ozone, carbon monoxide, sulphur oxides, nitrogen oxides, and lead. These pollutants can harm human health and the environment and cause property damage. The set of limits based on human health is called primary standards. Another set of limits intended to prevent environmental and property damage is called secondary standards [5].

The main pollutants measured to characterize and monitor the quality of the environment are shown below.

2.1. Particulate Matter (PM)

PM is one of the most important pollutants in terms of potential to harm human health, as it penetrates into low regions of the respiratory system. PM is a complex heterogeneous mixture of solid particles and liquid droplets found in the air, whose size and chemical composition change in time and space, depending on different emission sources and atmospheric and weather conditions. These particles come in many sizes and shapes and can be made up of hundreds of different chemicals. PM includes both primary and secondary PM; primary PM is the fraction of PM that is emitted directly into the atmosphere from construction sites, unpaved roads, fields, smokestacks, or fires, whereas secondary PM forms in the atmosphere following the oxidation and transformation of precursor gases (mainly SO_2, NO_X, NH3) and some volatile organic compounds (VOCs) that are emitted from power plants, industries and automobiles.

Particle pollution includes "inhalable coarse particles," with diameters larger than 2.5 μm and smaller than 10 μm, and "fine particles," with diameters that are 2.5 μm and smaller. Smaller sizes of PM such as $PM_{2.5}$, with a diameter up to 2.5 μm, are considered particularly harmful due to their greater ability to penetrate deep into the lungs.

2.2. Tropospheric or ground-level ozone (O_3)

Ozone (O_3) is a secondary pollutant formed in the troposphere from complex photochemical reactions following emissions of precursor gases such as NO_X and non-methane volatile organic compounds (NMVOCs), deriving from paint application, road transport, dry-cleaning, and other solvent uses. At the continental scale, methane (CH4) and carbon monoxide (CO) also play a role in O_3 formation. O_3 is a powerful and aggressive oxidizing agent, causing respiratory and cardiovascular problems and leading to premature mortality. High levels of O_3 can also damage plants, leading to reduced agricultural crop yields and decreased forest growth.

O_3 is found not only in the troposphere but also in the upper regions of the atmosphere. Three oxygen atoms constitute O_3, and it is a very reactive molecule. O_3 in the upper atmosphere protects the earth from the sun's harmful rays, whereas the ground level O_3 is considered as the main component of harmful smog.

In fact, the tropospheric O_3 is created by chemical reactions between oxides of nitrogen (NO_X) and volatile organic compounds (VOC) after its emission in the air from industrial facilities and electric utilities, motor vehicle exhaust, gasoline vapours, and chemical solvents. O_3 is likely to reach unhealthy levels on hot, sunny days in urban environments and can also be transported through long distances by wind so that even rural areas can experience high O_3 levels.

O_3 contributes to what we typically experience as "smog" or haze, which still occurs most frequently in the summertime but can occur throughout the year in some southern and mountain regions. Ground-level O_3 that we breathe can harm human health, especially affecting people with lung disease, children, older adults, and people who are active outdoors.

Children are at greatest risk from exposure to O_3 because their lungs are still developing and they are more likely to be active outdoors when O_3 levels are high, which increases their exposure. O_3 also damages vegetation, in particular, trees and plants during the growing season, and ecosystems, including forests, parks, wildlife refuges, and wilderness areas.

2.3. Carbon monoxide (CO)

Carbon monoxide (CO) is a very dangerous gas because it is odorless and it is emitted generally from all combustion processes both natural and anthropogenic sources. Urban air pollution is characterized by high CO emissions to ambient air from vehicles movement. CO can cause several harmful health effects, all related to reduced oxygen delivery to the body's organs (especially the heart and brain) and peripheral tissues causing death at very high doses of exposure.

2.4. Sulphur dioxide (SO_2)

Sulphur dioxide (SO_2) is emitted by the combustion of fuels containing sulphur. It contributes to acid deposition and can cause adverse effects on aquatic ecosystems in rivers and lakes, and damage to forests.

Sulphur dioxide (SO_2) is a highly reactive gas. SO_2 is emitted mainly by fossil fuel combustion by power plants (73%) and other industrial process (20%). Smaller sources of SO_2 are released by other industrial processes such as mineral extraction and also by the burning of high-sulphur fuels. Many illnesses are related to SO_2; particularly, harmful effects are linked to the respiratory system.

2.5. Nitrogen oxide (NO_X)

Nitrogen oxide (NO_X) is emitted during fuel combustion, by industrial facilities and the road transport. NO_X contributes to the formation of secondary inorganic PM and tropospheric O_3, to acid deposition and to eutrophication. Of the chemical species that comprise NO_X, NO_2 is associated with adverse affects on health, such as inflammation of the airways and reduced lung function.

Nitrogen dioxide (NO_2) is a highly reactive gas. Other nitrogen oxides include the nitrous acid and nitric acid. EPA's National Ambient Air Quality Standard uses NO_2 as indicator for the class of nitrogen oxide. NO_2 is emitted from all vehicles and industrial plants, including power plants. In addition to the arising O_3 and PM, NO_2 is linked also to several adverse effects on the respiratory system. Both primary EPA standard (to protect health) and secondary EPA standard (to protect the public welfare) for NO_2 are established at 0.053 parts per million (53 ppb), averaged annually. EPA (January 2010) established an additional primary standard at 100 ppb. Primary standards are finalized to protect public health, including health of responsive populations, people with COPD, children, and the elderly. No European and extra-European countries have been found stranger to the problem.

2.6. Heavy metals

The main heavy metals in outdoor air are arsenic (As), cadmium (Cd), lead (Pb), mercury (Hg), and nickel (Ni); they are emitted mainly as a result of various combustion processes and industrial activities and generally can reside in or be attached to PM. Heavy metals are persistent in the environment and can be deposited on terrestrial or water surfaces, contaminating soils or sediments, and accumulating in food-chains.

2.7. Polycyclic Aromatic Hydrocarbons (PAHs) and benzo(a)pyrene (BaP)

PAHs are pollutants created during the incomplete burning of several organic substances like coal, oil, gas, wood, garbage, and also tobacco and smoked and grilled. Actually, there are more than 100 different PAHs identified. PAHs generally occur as complex mixtures of congeners. The primary sources of exposure to PAHs are inhalation of the compounds in tobacco smoke, wood smoke, and ambient air.

PAHs is a growing health concern in Europe, with particular regard to BaP, a polycyclic aromatic hydrocarbon, formed mainly from the incomplete burning of organic material such as wood and from car exhaust fumes especially from diesel vehicles. It is a known cancer-causing agent in humans. A main source of BaP in Europe is domestic home heating, in particular, wood burning, waste burning, coke and steel production, and mobile sources. Other sources include outdoor burning and rubber tyre wear. In Europe, BaP pollution is mainly a problem in certain areas such as western Poland, the Czech Republic, and Austria where domestic coal and wood burning is common.

Between the PAHs, the International Agency for Research on Cancer (IARC) considers BaP as 2A class, so carcinogen. However, the carcinogenicity of BaP was demonstrated through studies carried out with animals. Benzo(a)pyrene is a known mutagen, but it must be metabolized to be able to induce mutation. Exposure to BaP pollution is quite significant and ubiquitous. Central and Eastern European countries report air BaP concentrations above the target value. Approximately one quarter of the EU people were exposed to higher BaP concentrations respect to target value (1 ng/m^3) for the period 2008–2010. But almost the entire population was exposed to BaP concentrations above WHO reference level [6]. The increase of BaP emissions in Europe over the last years is therefore an important concern for public health. Regarding BaP, the Industrial Emissions Directive [7] regulates emissions from a large range of industrial sources.

2.8. BTEX

It is a group of chemical compounds including Benzene, Toluene, Ethylbenzene, and Xylenes. BTEXs are made up of naturally occurring chemicals that are found mainly in petroleum products such as gasoline. Everyone is exposed to small amounts of BTEX compounds in the outdoor air, at work, and in their homes. Besides the common daily exposures to BTEX, larger amounts can enter the environment from leaking underground storage tanks, overfills of storage tanks, fuel spills from auto accidents, and landfills.

Short-term exposure to gasoline and its components benzene, toluene, and xylenes has been associated with skin and sensory irritation, central nervous system (CNS) problems, and effects on the respiratory system. Chronic exposure to BTEX compounds can affect the kidney, liver, and blood systems. Long-term exposure to high levels of the benzene compound can lead to leukemia and cancers of the blood-forming organs [8].

2.9. Physical pollution

Physical pollution is pollution caused by colour (change), suspended solids, foaming, temperature conditions, or radioactivity, and it is characterized by its influence on environmental conditions caused by forces and operations of physics, such as noise, microwave radiation, vibration.

2.10. Microbiological factors

Microbiological factors in the environment represent an underestimated but insidious risk factor, which concern has increased with the introduction of advanced technologies in hospitals, industry, and agriculture. Microbial agents are transported and diffused in ambient through PM [9]. The main sources of microorganisms are Flügge droplets from human airways and conditioning systems, the last containing different bacteria such as *Legionella pneumophila*, virus, and mould.

To maintain a healthy environment, a strict monitoring must be undertaken for each of the contaminants mentioned above; legislation on environment sets a benchmark and provides a series of actions to hold the values of the contaminants within the limits.

As air pollutants can be transported many miles away from the area of their emission, monitoring these substances is more difficult than monitoring contaminants in water or soil, so it was decided to record ambient air quality data focusing on major pollutants. Continuously operating automatic analysers are the most important information source on air pollution levels [10]. According to current legislation, measurements are taken at permanent sites equipped with physico-chemical monitoring devices even if these stations provide only data on exposure levels which do not facilitate making conclusions on the effects or impacts to human health and the general environment. Moreover, spatial coverage and temporal definition of the carried measurements together with meteorological conditions influence the degree of representativeness of the data compared to the real state of pollution of the study area [11].

Recently, innovative technologies have been developed and direct damage of harmful effects can be demonstrated by monitoring living organisms, called biomonitors, which provide a measure of integrated exposure over a certain amount of time taking into account also climatic conditions and enrich the substance to be determined so that the analytical accessibility is improved and the measurement uncertainty reduced. Moreover, sampling is relatively simple and no expensive technical equipment is needed. The field of biomonitoring includes bioindication of effects and bioaccumulation of pollutants [12].

The information provided by mosses, lichens, and higher plants on the deposition on the effects of air pollutants is an important complement to the data acquired with automatic systems. Distribution of tree species on a national or supranational lets you draw cheaply and in a short time maps of diffusion and deposition of persistent pollutants or the effects of tropospheric O_3 and other phytotoxic pollutants [13].

3. Statistical methodologies to evaluate the risk assessment and air pollution correlated diseases

Extreme pollution episodes that took place in the period 1930–1960 have initiated the epidemiological studies for increase of health environmental concern. The association between air pollution and certain health variables was made clear by several studies [14, 15].

Thanks to the new approach, air pollution levels were reduced substantially, such that, for health effects assessment, now, longer monitoring plans are required. To this end, epidemiologists began to use some statistical models. In the 1970s firstly were used the dynamic regression models, models in which the relationship between the dependent and explanatory variables were distributed over time rather than being expected to occur simultaneously. However, the problem of these types of models is that they assume that the dependent variable is distributed normally, but this is a condition extremely rare to achieve [16].

In the 1990s were used for the first time the linear models based on Poisson regression because the event counts more typically have a Poisson distribution. These models use the "time" and its transforms as variable [16].

Also Poisson regression is particularly useful only when cases rather than the entire population can be enumerated.

Nevertheless, Poisson regression poses the problem that, if any of these unmeasured variables follows a cyclical component of varying frequency, the parametric functions of time cannot be easily adapted. So, these limitations led to the development also of nonparametric Poisson regression [17], which is well adapted to the irregular cyclic components of unmeasured variables reducing any potential confounding.

One difficulty with this method is that the researcher must specify the number of degrees of freedom, with several discrepancies arising as the way to calculate this. Because inappropriate determination of the number of degrees of freedom is frequently a bias in the estimates of nonparametric Poisson designs, epidemiologists focused on the case-crossover (CCO) [18] design that purported to control time trends. The CCO design is characterized by the fact that each subject serves as his or her own control. This design was initially used to assess the effect of exposures measured at an individual level and was not applicable to exposures with a time trend, such as air pollution. A variant was developed to bypass this bias [19], the bidirectional CCO, characterized by having control time periods before and after the event. This design was already appropriate for ecologic-type exposures, such as air pollution, water pollution, etc. In addition, pollution values are not affected by the presence of prior morbidity and mortality events. In fact, in the CCO design in air pollution, pre- and post-event exposure values are independent of the hazard-period exposure; those that are post-event referent can be appropriate. One advantage of CCO design respect to Poisson regression is its ability to assess potential effect modification at the individual level rather than at the group level [20]. As an alternative analytic methodology to Poisson regression, the CCO approach allows for direct modeling of interaction terms, rather than depending on multiple subgroup analyses [20].

Besides the individual characteristics during the statistical analysis of epidemiological data, other variables must be considered, taking into account that the chemical characteristics of air pollutants mixture and PM may change over time and depending on the geographical location, emission sources, atmospheric chemistry, and weather conditions [21].

Interest in health effects of air pollution became more intense after two US cohort studies suggested that exposure to fine PM in the air was associated with life shortening [22,23].

Exposure to pollutants such as airborne PM and O_3 has been associated with increases in mortality and hospital admissions mainly because of respiratory and cardiovascular disease, due to both acute and chronic exposure [24].

Health problems can include cancer, respiratory irritation, nervous system problems, and birth defects.

Air pollution, in the year 2012, was responsible for 3.7 million deaths, i.e., approximately 6.7% of the total deaths. In particular, air pollution is estimated to cause deaths from lung cancer (16%), COPD (11%), ischemic heart disease and stroke (>20%), and respiratory infection (13%). Furthermore, the IARC announced that it has classified outdoor air pollution as carcinogenic to humans (Group 1).

Typically, exposure to toxic components of air pollutant causes the well-established correlated diseases, respiratory and pulmonary ones, such as decreased lung function and increased incidence of chronic cough, bronchitis, chronic obstructive pulmonary disease, asthma, and conjunctivitis [25,26,27].

Recent clinical and epidemiological data suggest also that cardiovascular disease may be related to pollution [28,29] especially those associates with fine PM [30,31,32,33]. The PM's effects on the cardiovascular system seem to involve the activation of prothrombotic factors, leading to thrombosis, but destabilization of atherosclerotic plaques cannot also be excluded. In addition, there may be direct effects on the heart or indirectly on the nervous system.

Health effects have been seen at very low levels of exposure, and it is unclear whether a threshold concentration exists for PM and O_3 below which no effects on health are likely.

It could be expected that the impact caused by a preventable risk factor would decline if the exposure to that risk factor could be reduced or removed. According to this approach, the proportional reduction in the number of health problems or deaths as a result of reducing the risk factor is known as the attributable fraction (AF) [34].

Public health agencies concerned with air quality perform risk assessments to determine the increased risk of illness from a specific human exposure to a toxic air pollutant.

The risk assessment approach outlined by the WHO in the Environmental Burden of Disease (EBD) series [35,36] includes the following steps:

1. assessment of the air exposure of the population through data from Air model or monitoring networks. A target concentration is also needed to determine the attributable disease or the potential gains of a management plan;
2. sufficient number of persons exposed to air pollutants;
3. baseline data of incidence of the adverse health outcomes associated with air pollutants (like mortality rate); and
4. concentration-response functions (CRFs) related to the incidence of adverse health effects.

The US Environmental Protection Agency (US-EPA) divided the risk assessment process into the following four steps:

a. **Hazard Identification**: it allows determining the potential human health effects from exposure to a chemical. This is based on information provided by the scientific literature.

b. **Dose-Response Assessment:** it is the assessment of the relationship between a dose of chemical exposure and incidence or severity of the related adverse health effect. It takes into consideration the intensity and pattern of exposure and also age and lifestyle variables that may affect people's susceptibility. The dose-response relationship is evaluated differently for carcinogenic and non-carcinogenic pollutants. In fact, for carcinogens, it is assumed that there is a linear relationship between an increased dose of exposure and increase in cancer risk; this is expressed as slope factor (SF) and threshold is not accepted. For risks evaluation by inhalation of carcinogenic substances, US-EPA use the potency slopes to develop the unit risk factors (URFs). A URF is the upper-bound excess probability of contracting cancer as the result of a lifetime of exposure to a carcinogen at a concentration of 1 $\mu g/m^3$ in air. For inhalation effects from non-carcinogens, dose-response data are used to develop reference concentrations (RfCs) for both long-term and short-term exposures. Unlike carcinogens, non-carcinogens are assumed to have thresholds dose, so the injury does not occur until exposure has exceeded a threshold limit. An RfC is derived from a no-observed adverse effect level (NOAEL) or lowest observed adverse effect level (LOAEL) determined through human or animal exposure studies.

c. **Exposure Assessment**: it determines the extent (intensity, frequency, and duration, or dose) of human exposure to a chemical in the environment.

d. There are three components to exposure assessment:

   i. Estimation of the maximum quantity of each pollutant emitted from the source.

   ii. For each contaminant emitted, estimation of the resulting maximum annual average and (where applicable) maximum short-term average ambient air concentrations, using dispersion models, or air impact values based on dispersion models.

   iii. Estimation of the amount of contaminant taken in by a human receptor.

e. **Risk Characterization**: it is the final step in risk assessment where health risk is calculated and described thanks to data collected in the first three steps.

- **Carcinogens**

Human health risk estimates for inhalation of carcinogenic air pollutants are based on the following:

$$\text{Cancer Risk} = C \times URF$$

where

C = maximum annual average ambient air concentration of a pollutant ($\mu g/m^3$)

URF = pollutant-specific inhalation unit risk factor $(\mu g/m^3)^{-1}$

For routes of exposure other than inhalation, risk is calculated by multiplying the estimated chemical dose (in mg/kg/day) by the chemical-specific oral slope factor (in $(mg/kg/day)^{-1}$).

- **No carcinogens**

Human health risk estimates for inhalation of non-carcinogenic air pollutants are based on the following:

$$\text{Hazard Quotient} = C/RfC$$

where

C = maximum ambient air concentration, $\mu g/m^3$

RfC = pollutant-specific reference concentration, $\mu g/m^3$

The averaging time can be either annual, or a specific number of hours, depending on the basis of the reference dose [37]. For routes of exposure other than inhalation, the hazard quotient is calculated by ratio between the estimated chemical dose (in mg/kg/day) and the chemical-specific reference dose (in mg/kg/day).

Hazard quotients can be summed (separately for inhalation and oral exposures and for different averaging times) to give a hazard index.

4. Evaluation of indoor air pollution and comparison with the urban and high-risk sites

Clean air is a fundamental health determinant. The quality of indoor air (e.g., homes, offices, schools, day care centres, public buildings, where people spend a large part of their life). Hazardous substances emitted from buildings, construction materials, and indoor equipment or due to all human activities indoors lead to a broad range of health problems and may even be fatal for infants and children. Indoor environmental issues are still an open topic of public health, including health risks and the means by which human exposures can be reduced [38].

Mistakes of evaluation or unacknowledged by public about health risks associated with a variety of indoor environmental pollutants and sources of pollution (e.g., radon, mold and moisture, secondhand smoke, and indoor wood smoke) cause a minor efficacy of prevention and/or risk's low perception. In poorly ventilated dwellings, smoke in and around the house can exceed acceptable levels for fine particles 100-fold [39] with respect to outdoor levels already known. In fact, inadequate ventilation can increase indoor pollutant levels for incomplete dilution of these. High temperature and humidity levels can also increase concentrations of certain pollutants [40]. Outdoor air enters and leaves a house by infiltration, natural ventilation, and mechanical ventilation. If infiltration, natural ventilation, or mechanical ventilation is poor, the air exchange rate is low and pollutant levels can increase rapidly [41].

There are several sources of indoor air pollution including combustion of oil, gas, kerosene, coal, wood, and tobacco products; building materials and furnishings as diverse as deteriorated, asbestos-containing insulation, wet or damp carpet, cabinetry and objects made with certain pressed wood products (e.g., formaldehyde, pesticides, etc.); products for household cleaning and their maintenance (phtalates, bisphenol A, fragrances, pesticides, detergents, chlorine bleach, lye, ammonia, etc.), cosmetics for personal care (solvents, phtalates, bisphenol A, fragrances, mineral oil, ethoxylated surfactants and 1,4-dioxane, formaldehyde, lead, oxybenzone, parabens, toluene, triclosan, etc.), chemicals used for hobbies (solvents, lead, pesticides, mineral oil, cadmium, manganese dioxide, cobalt, formaldehyde, aromatic and chlorinated hydrocarbons, ethylene glycol monomethyl ether, ethylene glycol monobutyl ether, etc); central heating and cooling systems and humidification devices (mold, viruses, fungi, bacteria, mycotoxins, etc.); and outdoor sources such as radon, pesticides, and outdoor air pollution likely to come into the house where can concentrate. Some of these sources (building materials, outdoor sources, wood furnishings, etc.) release pollutants more or less continuously. Other sources, related to activities carried out in the home, release pollutants intermittently, and their concentrations can remain in the domestic air for long periods after these activities. The relative importance of any single source depends on how much of a given pollutant it emits and by the proven hazard of those emissions. In some cases, factors such as how old the source is and whether it is properly maintained are very significant (e.g., an improperly functioning gas stove can emit more CO than one that is properly adjusted).

Exposure to indoor air pollution is particularly high among women and young children who spend the most time in the domestic environment, and health effects may be shown soon after exposure or years later.

According to WHO, 4.3 million people a year die from the exposure to household air pollution [39]. In fact, especially cooking and heating with solid fuels (wood, charcoal, etc.) produce high levels of smoke both in and around the home and this contains a large variety of health-damaging pollutants [42].

Immediate effects such as irritation of the eyes, nose, and throat; headaches; dizziness; and fatigue may show up after a single acute exposure or repeated sub acute exposures. These are usually treatable and have short-term effects. Oftentimes, the simple elimination of a person's exposure to the identified source of pollution is sufficient for restoring health.

An additional problem is the recurrent multiple indoor pollution exposure of residents, complicating and amplifying the health effects [43].

Exposure to indoor air pollution leads to a wide range of respiratory diseases [39, 44] such as lung cancer [45, 46], ischemic heart disease [46], stroke [46], and cataract [46].

There is emerging evidence, although based on fewer studies, that suggests that household air pollution, especially in developing countries, may also increase the risk of other important child and adult illnesses such as the following:

- low birth weight and perinatal mortality (still births and deaths in the first week of life) [45, 46],
- asthma [39, 45],
- otitis media (middle ear infection) and other acute upper respiratory infections [39],
- tuberculosis [45],
- nasopharyngeal cancer [39],
- laryngeal cancer [39], and
- cervical cancer [39].

Also exposure to unhealthy concentrations of fine PM has been connected to increased respiratory/cardiovascular illnesses [39, 47]. In fact, the smaller air particles can penetrate into the deeper lung. PM is being linked to adverse birth outcomes [46], neurodevelopment, cognitive function [48], and diabetes [39].

Semple et al. [49] said that median $PM_{2.5}$ concentrations from 93 smoking homes were 31 $\mu g/m^3$ (ranged between 10 and 111 $\mu g/m^3$) and 3 $\mu g/m^3$ (ranged between 2 and 6.5) $\mu g/m^3$ for the 17 non-smoking homes and still showed that non-smokers living with smokers typically have average $PM_{2.5}$ exposure levels more than three times higher than the WHO guidance for annual exposure to $PM_{2.5}$ (10 $\mu g/m^3$). So, $PM_{2.5}$ pollution in indoor ambient where smokers live is approximately 10 times higher than that of non-smoking. A non-smoker living with a smoker is exposed to same $PM_{2.5}$ of a non-smoker living in a heavily polluted city such as Beijing. This condition is likely to be greatest and dangerous for the very young and for older members of the population because they typically spend more time at home. Zhou and colleagues [50] have demonstrated that in New York City, despite the ban on tobacco smoking, some hookah bars still serve tobacco-based hookahs, and in these particular indoor environments, the authors have found elevated concentrations of pollutants that may present a real health threat to visitors and employees. The mean real-time $PM_{2.5}$ level was 1179.9 $\mu g/m^3$, whereas the filter-based total PM mean was 691.3 $\mu g/m^3$. The mean real-time black carbon level was 4.1 $\mu g/m^3$, organic carbon was 237.9 μg/m^3, and CO was 32 ppm. Airborne nicotine was present in all studied hookah bars (4.2 μg/m^3).

Dorizas et al. [51] with their study showed that in nine naturally ventilated primary schools of Athens (Greece) during spring, PM concentrations were significantly affected by the ventilation rates and presence of students. Both PM_{10} and $PM_{2.5}$ were greater during teaching than the non-teaching hours, and, in many cases, the $PM_{2.5}$ concentrations exceeded their limit values. For most of the cases, the indoor to outdoor concentrations ratios of PM_{10} and $PM_{2.5}$ were much greater than one, indicating that the indoor environment was being mostly affected by indoor sources instead of the outdoor air. Furthermore, it was found that chalk and marker board usage significantly affects indoor pollutant concentrations.

No similar characteristics were found between indoor pollution and urban and high-risk sites outdoor pollution, because the poor ventilation of houses and buildings in general allows the concentration of chemical and biological pollutants that are not found in similar outdoor concentrations also in severe pollution event. Urban outdoor air pollution refers to the air pollution which the populations are exposed to, living in and around the urban area. Indoor air pollution refers to the pollutants found in indoors. An important difference is in the heavy metals concentration, generally most abundant and with major variability of species in urban and industrial air outdoor pollution compared with indoor pollution.

At the moment, in Italy, a reference rule has not been set. For this reason, until today, the main information concerning some guidelines or reference values in indoor air is obtained by the international scientific literature or by the few guidelines issued by other European countries or, for analogy, by other guidelines values regarding outdoor air. However, public health awareness on indoor air quality still lags significantly behind that of outdoor air quality.

The main areas at high risk of environmental crisis in Italy are described below.

5. Case reports: ILVA-Taranto, Melilli-Priolo-Augusta sites, Gela and Milazzo emerging situation, "Terra dei Fuochi" and Seveso

In order to cope with the huge problem of pollution, experts were mobilized at national and regional levels in the drafting of numerous scientific papers in order to initiate actions to protect health in the areas most affected by the emission of high quantities of contaminants. In many cases, the results of these studies support arguments about correlation between the exposition to pollutants and outbreak of neoplastic diseases; however, for some areas, it has been highlighted that there is a higher incidence in urban areas rather than in sites of national interest, leading to speculation about the etiologic implication of non-industrial contaminants.

At the moment in Italy, the industrial pole of Taranto, Melilli-Priolo-Augusta, Gela e Milazzo sites, and "Terra dei fuochi" and the outcome of the tragedy that occurred in 1976 in Seveso represent the most interesting areas for their extension and extent of involved population. An overview of current developments of studies and the status of remediation undertaken in these areas will be given below.

5.1. ILVA-Taranto

Founded in 1961, ILVA of Taranto is a steel plant at full cycle, where occur all the steps that lead from iron ore to steel [52–54]. The plant, which is partially impounded by order of the judiciary, is the largest steel maker in Europe. The area of Taranto is identified at high risk of environmental crisis because of an extensive industrial area developed close to the urban settlement. Industrial activities are responsible for environmental pollution, mostly due to polycyclic aromatic hydrocarbons (PAHs), heavy metals, organic solvents, polychlorinated biphenyls (PCBs), PM, and dioxin.

Table 1 shows the main pollutants found in soil, ground waters, and sediments of the area of Taranto.

Soil	antimony, arsenic, beryllium, cadmium, cobalt, chromium total, chromium VI, mercury, lead, nickel, zinc, copper, vanadium, cyanide, hydrocarbons C <12 and C"/> 12, PAHs, benzene, xylene, dioxins
Subsurface	antimony, arsenic, beryllium, cadmium, cobalt, chromium total, chromium VI, mercury, lead, nickel, zinc, copper, vanadium, cyanide, hydrocarbons C <12 and C"/> 12, PAHs, benzene, xylene, dioxins
Groundwater	arsenic, selenium, aluminium, iron, manganese, nickel, lead, cobalt, total chromium, chromium VI, cyanides, sulphates, nitrates, BTEXS, PAHs, hydrocarbons and MTBE
Sediments	arsenic, nickel, lead, total chromium, copper, mercury, zinc, total PAHs, PCBs

Table 1. Main pollutants found in soil, ground waters, and sediments of the area of Taranto

To define the extent of the health risk, rigorous epidemiological studies have been conducted to assess both the short-term effects that occur in acute as a result of rapid changes in the concentrations of pollutants and the long-term effects caused by prolonged exposures and which occur 10–15 years after the start of exposure.

In 2006, the Ministry of Health has funded a project called "SENTIERI" (Studio Epidemiologico Nazionale dei Territori e degli Insediamenti Esposti a Rischio da Inquinamento - National Epidemiologic Study of the Territories and Settlements Exposed to Risk from Pollution) with the purpose to analyse the mortality of populations residing in proximity of a number of industrial agglomerates which by their nature potentially have a high factor of hazardous health and/or environmental contamination such as to be classified as SIN (Sites of National Interest for the Remediation).

The population of Taranto has been the subject of several multi-centre epidemiological studies that have documented the role of air pollution on the increase in short-term and long-term effects [52,53].

Epidemiological analysis of the residents of the city revealed death rates from all causes, for lung, pleura, and bladder cancer and for non-Hodgkin lymphoma. A case-control study on incident cases of these diseases in Taranto suggested a possible link between risks and residence close to the sources of emissions.

A recent study called EpiAir in which subjects were recruited rigorously and it was taken into account the exact georeferencing of areas of residence and employment analysed the spatial variability of air pollutants in Taranto and showed that atmospheric pollution in this city is not distributed homogeneously, but it is mainly widespread in the areas adjacent to industrial pole. In the results of this study have been taken into account the socio-economic status and calculated indices of deprivation; these aspects have a great importance, as many individual habits, such as cigarette smoking, alcohol consumption, physical activity, and obesity, are often associated with social status so that the adjustment made for the socio-economic factor has also adjusted the individual variables not directly measured, providing very reliable data. Moreover, in this study, mortality rate and hospitalization rate were measured to evaluate both long- and short-term effects. Data about mortality are reliable, as 98% of the causes of death were recovered, thanks to the linkage of personal data with the database of the ASL and show an excess of mortality for cancer, cardiovascular, and respiratory diseases. As far as hospital admissions, evaluated from hospital discharge schedule, were concerned, it was difficult to get accurate data because the comparison was made only on the hospital area of Taranto, without taking into account the extra-regional mobility and the secondary diagnosis. Although these are limited, data show an increase of hospitalization for the abovementioned causes. The results of this study have strengthened the argument that there is a correlation between exposure to emissions from the steel plant of Taranto and an increase of cancer and cardiovascular and respiratory diseases [55].

The European Environment Agency, in the list that shows the 622 most polluting industrial plants in Europe, included more than 60 Italian companies; Ilva of Taranto is placed second. As a result, the situation with the city of Taranto and Ilva is currently the subject of discussion and great concern [53,54,56]. To control the situation in the area were enacted several regional laws aiming to carry out the assessment of environmental impact and damage health in order to take targeted action to protect the environment and public health. Moreover, in the Italian Act 6 of February 2014, strategies are defined to make the remediation of contaminated areas from emissions of ILVA.

5.2. Sicily air quality

In Sicily, four Sites of National Interest have been individuated and the industrial areas of Melili-Priolo-Augusta (Siracusa), Gela (Caltanissetta), and Milazzo (Messina) have been declared by national and regional legislation "areas high risk of environmental crisis." It is likely that the excesses of mortality and morbidity observed in areas of Melili-Priolo-Augusta, Gela, and Milazzo are attributable to occupational exposures and environmental concerns related to the number of plants and the consequent contamination of environmental matrices [57], although the last reports show how these rates do not exhibit a significant increase.

5.2.1. Melilli-Priolo-Augusta sites

The large industrialized coastal area of eastern Sicily within the territory of the municipalities of Augusta, Priolo Gargallo, and Melilli is defined “petrochemical pole of Siracusa.” These territories started to be subject to industrialization in 1948 with the construction of a refinery (the RA.SI.O.M), and in the following years, other industrial facilities were born; currently, the petrochemical pole hosts five refineries of petroleum products, two centrals of ENEL, a gas plant and cogeneration, a factory of magnesite, a cement plant, a purifier of industrial and urban waste and a shipyard.

In these areas, the increase in the number of illness might be linked with PM, which are released into the atmosphere from industrial chimneys and have the ability to convey inside the body, through the lung, every kind of pollutants from the atmosphere. The residents, who have expressed their concerns to environment organizations and public heath institutions, have also noted this potential association.

Official data on emissions point out a long list of definitely carcinogenic and teratogenic chemicals (acrylonitrile, benzene, cadmium, hexavalent chromium, nickel, silicon, vanadium, dioxins, and furans), and many other potential hazards, according to the IARC (Agency International Agency for Research on Cancer).

About PM, the main substances contained in it are represented by heavy metals as shown by the results of a study conducted by the Faculty of Agriculture of the University of Palermo, which used lichen for biomonitoring as their action of bio-accumulators for heavy metals, highlighting a significant presence of these hazardous metals in the environment concerned [57, 58].

These pieces of evidence induced the Ministry of the Environment to recognize the Priolo-Augusta-Melilli site as "an area in environmental crisis for which it has become necessary a legislative and Financial able to address in a timely manner the dramatic emergency." The large amount of epidemiologic studies conducted in these areas highlights the role of atmospheric pollution in the development of some cancers although recent data published by AIRTUM (Associazione Italiana dei Registri Tumori – Italian Association of cancer registries) [59] show that the higher incidence of cancer in Sicily is found in the city of Catania and in other metropolitan areas, such as Palermo and Messina, as well as in big countries like Acireale and Gravina, leading to hypothesize the role of other risk factors probably linked to road traffic and unhealthy lifestyle, contributing in the outbreak of above mentioned diseases. According to the last published report, the industrial sites are not among the areas with the highest incidence of cancer, as supposed, and surprisingly, the SIN of San Filippo del Mela does not present excess in the incidence of tumors.

5.2.2. Gela and Milazzo emerging situation

In the area of Gela and the surrounding area rises one of the main industrial centres of the island. The area, declared "area at high risk of environmental crisis," includes the municipalities of Gela, Niscemi, and Butera. Within the bounded area subjected to remediation, there is a large industrial centre, consisting of plants for refining and extraction of crude oil and petrochemical plants. In particular, productions include polyethylene, molten sulphur, sulphuric acid and phosphoric acid, ammonia and fertilizer complexes [57].

The environmental impact of the refinery of Gela comes mainly from emissions into the atmosphere and the consequent presence of pollutants such as SO_2, NO_X, dust, H_2S, and CO. Conveyed emissions are mainly generated in the combustion process and are coming out of the chimneys.

Data provided by ARPA (Regional Agency for Environmental Protection) reveal exceedances in ambient town, some parameters such as benzene, methane, hydrocarbons, and PM_{10} containing heavy metals such as nickel. A scientific study has shown that the entire area of Gela is heavily influenced by emissions of metals and non-metals from both car traffic or vehicular pollution and industrial pollution. In particular, the study highlights pollution of the particulate present in the aerosol analysing the deposition of the particulates on pine needles. The analysis of pine needles is a fast method for monitoring the pollutants present in the air: the pine needles provide information for the long-term impact of even low levels of pollutants. As for neoplastic diseases, it has not been shown that there is an excess of incidence in this area, although some Sicilian studies highlight an increase of Relative Risk for these diseases compared to other areas of the Region not characterized by the presence of industrial agglomerates. Moreover, in this area, an important role is played by the current wind, responsible for the transport of substances emitted by plants in Gela for long distances, interestingly in particular to Niscemi where contamination of soils by the above mentioned pollutants was found.

The industrial area of Milazzo is characterized by the presence of a large industrial centre, including an oil refinery, a thermoelectric central, a co-generator, a steel mill, a plant for the recovery of lead from spent batteries, and different other smaller factories. In the vicinity of industrial installations, there are several common installations, among which are those belonging to "area at high risk of crisis environment."

A single study carried out in Milazzo area reveals that children are exposed to elevated concentrations of sulphur dioxide, nitric oxide, and $PM_{2.5}$. In more weeks, the average concentration evaluated for sulphur dioxide has exceeded 20 $\mu g/m^3$, and the $PM_{2.5}$ average in the study period was about 23 $\mu g/m^3$, i.e., twice the WHO air guideline limit. Also, epigenetic markers were associated with air pollutant concentrations – particularly with regard to the nitric oxide pathway [57, 58].

5.3. Terra dei fuochi

One of the greatest environmental and health criticality that covered Italy in the last two decades is linked to the illegal waste disposal, triggering serious repercussions both in the territories concerned and the health of the residents in these areas and leading the Institutions to take actions that stiffen confidence in themselves and protect the health of the population. Studies carried out in the areas affected by this phenomenon, although not definitive, confirmed an increase in health risks resulting from waste disposal, highlighting in particular an increase in mortality from all causes, the excesses of mortality from specific cancers (liver, lung, stomach, kidney, and bladder) as well as an increase in non-neoplastic diseases of the respiratory system. The limits of ecological studies mainly reside in the fact that it is not possible to assess the individual characteristics of involved subjects, data on the amount and types of substance emitted are not exact, and therefore, extent of population exposure is not certain and estimates arising from these studies do not provide reliable data on the correlation. Despite the necessity of deepening the studies above, on February 2014, the Italian Parliament approved a law for the protection of these specific areas, which has as main innovations the request for a mapping of polluted areas and crops of these territories, the allocation of funds to carry out a health screening, and the introduction of the crime of burning waste.

This law, if one side has paid attention on the problem, on the other hand has created problems for the application of screening, excellent in itself as a means of prevention, but totally inappropriate to the case in question. It should be noted that on such land occurred the burning of hazardous waste, even radioactive in nature, and that substances emanating from these processes have not been proven to be in correlation with cancers for which screening is scheduled (colorectal, breast, cervix). Moreover, alarmism generated by media has also created skepticism in consumers, even internationally, with a substantial drop in sales of the excellent products from Campania.

However, contrasting the illegality and protecting public health is a duty of the institutions, as well as civic, and to implement the cited actions, the Institutions with the collaboration of experts gathered in the Technical Committee for the Contaminated Territories (CTTC) that provide information on correct methodology to implement the definition of the criteria for the mapping of sites, for the identification of specific diseases to perform a targeted screening, and for implementing the operations of monitoring and remediation.

5.4. Seveso

On July 1976 in Meda (Italy) at the chemical plant ICMESA (Industrie Chimiche Meda Società Azionaria) occurred the Seveso disaster [60]. During the usual production of trichlorophenol, a fungicide, an uncontrolled reaction that sparked the safety valves of the tank reactor took place, releasing the deadly chemical vapour into the atmosphere. About 3000 kg of toxic cloud compound containing dioxin TCDD (2,3,7,8-tetrachlorodibenzo-p-dioxin) was released into the air. The wind immediately dispersed the toxic cloud eastwards: the dioxin could begin to generate its catastrophic effects, affecting the inhabitants of the area and also making uninhabitable the territory on which it was deposited.

The chemicals released into the air from the ICMESA were carried by the wind and caused the contamination of, mainly, four municipalities: Seveso, Meda, Desio, and Cesano Maderno.

Since no one was at the plant when it happened, the seriousness of the disaster was not immediately observed.

Dioxin has been considered to be the most toxic human-made substance; the Seveso disaster represents the highest known TCDD exposure to residential populations and probably the most studied dioxin contamination incident in history [60, 61].

The most common sign of human health problems was skin disorder. The immediate effects on the population were evident especially from a dermatological point of view: after two days already appeared the first cases of chloracne, a disease that is documented to be associated with dioxin.

In order to estimate health problems of population, a great number of laboratory tests were carried out on blood samples. In 1987, it became possible to measure low levels of dioxin in blood samples. Another important effect of the dioxin exposure concerned reproduction, in particular affecting the sex ratio (a higher proportion of females were born in the first 7 years after the accident). In addition, a large number of pregnancies ended as spontaneous abortions were reported.

The affected places still show signs of contamination. The new EU legislation on industrial equipment of chemical and their coordinated management was started in June of 1982, when the Council of Ministers approved the European Union's Council Directive 82/501/EEC, the Seveso Directive.

A central part of the directive was made by reference to the obligation of transparent public information about industrial accidents, as well as new security measures to be taken in case of an accident [60, 61].

The new concept was approved for industrial workers and the public, the right to know the substance and the form of the problems that could threaten them and the exact operation of the safety procedures.

6. Conclusions

From the analysis of situation in the sites at risk of environmental crisis on the Italian territory, the greatest criticalities are represented by the effects of emissions from industrial plants although increasing attention must be paid to the effects of vehicular traffic, as shown by recent data of RTI report, according to which the highest incidence of neoplastic diseases is recorded in areas characterized by the presence of large urban agglomerations.

As regards the comparison between the role of indoor and outdoor pollution in the onset of diseases related to environmental exposure, the second is still predominant and, in the case of sites of national interest, indoor air pollution can be regarded as a factor that may enhance the effects induced by outdoor pollutants.

Pollution situation is controlled through the application of numerous laws on environmental protection, which in our country are very stringent and thanks to which it was possible at least to contain both the emission of pollutants and their effects on human health.

The effect of environmental stressors on human health and policy-making today can be explained using the Driving Force-Pressure-State-Exposure-Effect-Action (DPSEEA) model [62]. Ideal environmental management is based on a reduction of environmental risk exposure, as well as the risks themselves, in all stages of the process. However, this policy-based intervention is not accurate or simple for several motifs (problematic approach for many variables as time lag between exposure reduction to environmental stressor and its negative effects on health or, e.g., as many other variables) that make the intervention very difficult to implement [63]. For DPSSEA model application, the national policy-based intervention, finalized to air pollutants reduction, is very closely to the air quality standards, as the setting-up of those standards can lead to management and reduction of sources of pollution so decreasing pollutant concentrations and subsequent relevant health risk factors to ensure healthy residents [62, 64]. Moreover, air quality standards setting may be considered to be a reliable method for reducing exposure to air pollutants.

Although Institutions have implemented several containment measures, environmental pollution neither has nor ceased to be a problem of Public Health. Currently, in Italy, there are about 15,000 sites to characterize, reclaim, and monitor that represent more than 2% of the entire territory and require the allocation of 25–30 billion euros.

To solve this problem, firstly, rigorous epidemiologic studies must be previewed and conducted, taking into account a careful evaluation phase to confirm the association between environmental exposure and health effects. Epidemiological studies should provide for a careful characterization of the sites, taking into consideration that the great variety of production processes gives rise to an equally wide variety of pollutants not known and for which are not yet known metabolism and mechanisms of interaction with other substances present or released into the environment.

Performing a thorough epidemiological investigation is essential to proceed to risk assessment and risk management. In the management of this route, greater emphasis must be attributed to activities of prevention departments and ARPA and technical committees consisting of experts who can give useful indications to legislators to improve and promulgate laws to protect collective health and must perform a central role. The aim of this cooperation is achieving a "sustainable risk," that is, the risk that remains after the interventions of control and prevention and that can be more easily managed by the competent bodies.

Nonetheless, it is necessary to consider that the bases to make the most of these actions are the training of the operators and the information, so that in this view, activities of continuous updating for public health professionals are promoted.

Author details

M. Ferrante*, M. Fiore, C. Copat, S. Morina, C. Ledda, C. Mauceri and G. Oliveri Conti

*Address all correspondence to: marfer@unict.it

Environmental and Food Hygiene Laboratories (LIAA), Department of Medical Science, Surgical and Advanced Technologies "GF Ingrassia", Hygiene and Public Health, University of Catania, Italy

References

[1] World Health Organization (WHO). http://www.who.int/en/.

[2] European Environmental Agency (EEA). http://www.eea.europa.eu/.

[3] Commission for Environmental Cooperation (CEC). The North American Mosaic: An Overview of Key Environmental Issues. www.cec.org.

[4] European Environmental Agency (EEA). Technical report No 9/2009.

[5] United States Environmental Protection Agency (US EPA). http://www.epa.gov/air/urbanair/.

[6] European Environmental Agency (EEA). Air Quality in Europe – 2013 Report. Report No 9/2013. ISSN 1725-9177.

[7] Europe Union (EU). Directive 2010/75/EU of the European Parliament and of the Council of 24 November 2010 on Industrial Emissions (integrated pollution prevention and control). 2010; OJ L 334: 17–119.

[8] Agency for Toxic Substances and Disease Registry (ATSDR). Interaction profile for Benzene, Toluene, Ethylbenzene, and Xylenes (BTEX). 2004.

[9] Pasquarella C, Pitzurra O, Savino A. The index of microbial air contamination. *Journal of Hospital Infection*. 2000; 46: 241–256.

[10] Zauli Sajani S, Scotto F, Lauriola P, Galassi F, Montanari A. Urban air pollution monitoring and correlation properties between fixed-site stations. *J. Air & Waste Manage. Assoc.* 2004; 54:1-6. ISSN 1047-3289.

[11] Allegrini I, De Santis F, Perrino C, Pietrodangelo A. Metodi innovativi per il monitoraggio dell'inquinamento atmosferico. *Silvae, Anno III*. 8: 45-64.

[12] International Atomic Energy Agency (IAEA). Biomonitoring of Atmospheric Pollution (with Emphasis on Trace Elements) – BioMAP II. IAEA-TECDOC 2003. 1338. ISBN 92–0–100803–1. ISSN 1011-4289.

[13] Bargagli R. Monitoraggio degli inquinanti atmosferici persistenti mediante i muschi e le piante superiori. *Biologia Ambientale*. 2006; 20 (2): 55-67.

[14] Firket J. The cause of symptoms found in the Meuse Valley during the fog of December 1930. *Bull Acad R Med Belg*. 1931; 11:683–741.

[15] Logan WPD. Mortality in the London fog incident, 1952. *Lancet*. 1953; 1:336–338.

[16] Saez M, Perez-Hoyos S, Tobias A, Saurina C, Barceló MA, Ballester F. Métodos de series temporales en los studies epidemiológicos sobre contaminación atmosférica [in Spanish]. *Rev Esp Salud Publica*. 1999; 73:133–143.

[17] Kelsall JE, Samet JM, Zeger SL, Xu J. Air pollution and mortality in Philadelphia, 1974–1988. *Am J Epidemiol*. 1997; 146:750–762.

[18] Maclure M. The case-crossover design: a method for studying transient effects on the risk of acute events. *Am J Epidemiol*. 1991; 133:144–153.

[19] Navidi W, Thomas D, Langholz B, Stram D. Statistical methods for epidemiologic studies of the health effects of air pollution. *Res Rep Health Eff Inst*. 1999; 86: 1–50.

[20] Figueiras A, Carracedo-Martinez E, Saez M, Taracido M. Analysis of case-crossover designs using longitudinal approaches: a simulation study. *Epidemiology*. 2005; 16:239–246.

[21] WHO. Health relevance of particulate matter from various sources. Report on a WHO Workshop. Bonn, Germany, 26-27 March 2007.

[22] Dockery DW, Pope CA 3rd, Xu X, *et al.* An association between air pollution and mortality in six US cities. *N Engl J Med*. 1993; 329:1753–1759.

[23] Pope CA 3rd, Thun MJ, Namboodiri, MM *et al.* Particulate air pollution as a predictor of mortality in a prospective study of US adults. *Am J Respir Crit Care Med*. 1995; 151:669–674.

[24] Brunekreef B, Holgate ST. Air pollution and health. *The Lancet*. 2002; 360(9341):1233–1242.

[25] Dockery DW, Pope CA 3rd, Xu X, Spengler JD, Ware JH, Fay ME, Ferris BG Jr, Speizer FE. An association between air pollution and mortality in six U.S. cities. *N Engl J Med*. 1993; 329:1753-1759.

[26] Hoek G, Brunekreef B, Goldbohm S, Fischer P, Van den Brandt PA. Association between mortality and indicators of traffic-related air pollution in the Netherlands: a cohort study. *Lancet*. 2002; 360:1203-1209.

[27] Pope CA 3rd, Burnett RT, Thurston GD, Thun MJ, Calle EE, Krewski D, Godleski JJ. Cardiovascular mortality and long-term exposure to particulate air pollution: epidemiological evidence of general pathophysiological pathways of disease. Circulation 2004; 109:71-77.

[28] Brook RD, Franklin B, Cascio W, Hong Y, Howard G, Lipsett M, Luepker R, Mittleman M, Samet J, Smith SC Jr, Tager I. Air pollution and cardiovascular disease: a statement for healthcare professionals from the Expert Panel on Population and Prevention Science of the American Heart Association. *Circulation*. 2004; 109:2655-2671.

[29] Hoek G, Brunekreef B, Fischer P, van Wijnen J. The association between air pollution and heart failure, arrhythmia, embolism, thrombosis, and other cardiovascular causes of death in a time series study. *Epidemiology*. 2001; 12:355-357.

[30] Gold DR, Litonjua A, Schwartz J, Lovett E, Larson A, Nearing B, Allen G, Verrier M, Cherry R, Verrier R. Ambient pollution and heart rate variability. *Circulation*. 2000; 101:1267-1273.

[31] Wellenius GA, Schwartz J, Mittleman MA. Particulate air pollution and hospital admissions for congestive heart failure in seven United States cities. *Am J Cardiol*. 2006; 97:404-408.

[32] Zanobetti A, Schwartz J. The effect of particulate air pollution on emergency admissions for myocardial infarction: a multicity case-crossover analysis. *Environ Health Perspect*. 2005; 113:978-982.

[33] Pope CA 3rd, Muhlestein JB, May HT, Renlund DG, Anderson JL, Horne BD. Ischemic heart disease events triggered by short-term exposure to fine particulate air pollution. *Circulation*. 2006; 114:2443-2448.

[34] Prüss-Üstün A, Corvalán C. Preventing disease through healthy environments. Towards an estimate of the environmental burden of disease. WHO, 2006. http://www.who.int/quantifying_ehimpacts/ publications/preventingdisease/en/.

[35] Ostro B. Outdoor air pollution: assessing the environmental burden of disease at national and local levels. Geneva: World Health Organization. 2004. http://www.who.int/quantifying_ehimpacts/publications/ebd5/en/.

[36] Prüss-Üstün A, Mathers C, Corvalán C, et al. Introduction and methods: assessing the environmental burden of disease at national and local levels. Geneva: World Health Organization, 2003. http://www.who.int/quantifying_ehimpacts/publications/9241546204/en/.

[37] US-EPA. Guidelines for Carcinogen Risk Assessment, March 2005.

[38] Rivas I, Viana M, Moreno T, Pandolfi M, Amato F, Reche C, Bouso L, Àlvarez-Pedrerol M, Alastuey A, Sunyer J, Querol X. Child exposure to indoor and outdoor air pollutants in schools in Barcelona, Spain. *Environ Int*. 2014 Aug; 69:200-12. doi: 10.1016/j.envint.2014.04.009.

[39] WHO. Indoor Air Quality Guidelines: selected pollutants. 2010. ISBN 978 92 890 0213 4

[40] Pegas PN, Alves CA, Evtyugina MG, Nunes T, Cerqueira M, Franchi M, Pio CA, Almeida SM, Freitas MC. Indoor air quality in elementary schools of Lisbon in spring. *Environ Geochem Health*. 2011 Oct; 33(5):455-68.

[41] Kim C, Gao YT, Xiang YB, Barone-Adesi F, Zhang Y, Hosgood HD, Ma S, Shu XO, Ji BT, Chow WH, Seow WJ, Bassig B, Cai Q, Zheng W, Rothman N, Lan Q. Home kitchen ventilation, cooking fuels, and lung cancer risk in a prospective cohort of never smoking women in Shanghai, China. *Int J Cancer*. 2014 Jun 11. doi: 10.1002/ijc.29020.

[42] Gordon SB, Bruce NG, Grigg J, Hibberd PL, Kurmi OP, Lam KB, Mortimer K, Asante KP, Balakrishnan K, Balmes J, Bar-Zeev N, Bates MN, Breysse PN, Buist S, Chen Z,

Havens D, Jack D, Jindal S, Kan H, Mehta S, Moschovis P, Naeher L, Patel A, Perez-Padilla R, Pope D, Rylance J, Semple S, Martin WJ. Respiratory risks from household air pollution in low and middle income countries. *Lancet Respir Med.* 2014 Oct; 2(10): 823-860. doi: 10.1016/S2213-2600 (14) 70168-7.

[43] Colton MD, MacNaughton P, Vallarino J, Kane J, Bennett-Fripp M, Spengler JD, Adamkiewicz G. Indoor air quality in green vs conventional multifamily low-income housing. *Environ Sci Technol.* 2014 Jul 15; 48(14):7833-41.

[44] Alam DS, Chowdhury MAH, Siddiquee AT, Ahmed S, Hossain MD, Pervin S, Streatfield K, Cravioto A, Niessen LW. Adult cardiopulmonary mortality and indoor air pollution. a 10-year retrospective cohort study in a low-income rural setting. *GLOBAL HEART.* 2012; 7:(3):215–e221.

[45] Fullerton DG, Bruce N, Gordon SB. Indoor air pollution from biomass fuel smoke is a major health concern in the developing world. *Transactions of the Royal Society of Tropical Medicine and Hygiene.* 2008; 102:843-851.

[46] Ghosh R. Indoor smoke and prenatal and childhood growth: the role of (gestational) age. *World J Clin Pediatr.* 2013 Nov 8; 2(4):31-5. doi: 10.5409/wjcp.v2.i4.31.

[47] Challoner A, Gill L. Indoor/outdoor air pollution relationships in ten commercial buildings: $PM_{2.5}$ and NO_2. *Building and Environment.* 2014; 80:159e173.

[48] Tzivian L, Winkler A, Dlugaj M, Schikowski T, Vossoughi M, Fuks K, Weinmayr G, Hoffmann B. Effect of long-term outdoor air pollution and noise on cognitive and psychological functions in adults. *Int J Hyg Environ Health.* 2014 Sep 3. pii: S1438-4639 (14) 00077-7. doi: 10.1016/j.ijheh.2014.08.002.

[49] Semple S, Apsley A, Azmina Ibrahim T, Turner SW, Cherrie JW. Fine particulate matter concentrations in smoking households: just how much secondhand smoke do you breathe in if you live with a smoker who smokes indoors? *Tob Control.* 2014 Oct 20. pii: tobaccocontrol-2014-051635. doi: 10.1136/tobaccocontrol-2014-051635.

[50] Zhou S, Weitzman M, Vilcassim R, Wilson J, Legrand N, Saunders E, Travers M, Chen LC, Peltier R, Gordon T. Air quality in New York City hookah bars. *Tob Control.* 2014 Sep 16. pii: tobaccocontrol-2014-051763. doi: 10.1136/tobaccocontrol-2014-051763.

[51] Dorizas PV, Assimakopoulos MN, Helmis C, Santamouris M. An integrated evaluation study of the ventilation rate, the exposure and the indoor air quality in naturally ventilated classrooms in the Mediterranean region during spring. *Sci Total Environ.* 2014 Oct 6; 502C: 557-570. doi: 10.1016/j.scitotenv.2014.09.060.

[52] Tonelli F, Short S.W., Taticchi P. Case study of ILVA, Italy: The impact of failing to consider sustainability as a driver of business model evolution. www.isss.it, Istituto Superiore Sanità, accessed 10/09/2014.

[53] Boffetta P, La Vecchia C, Lotti M, Moretto A. Commenti al documento "Valutazione del danno sanitario stabilimento ILVA di Taranto ai sensi della LR21/2012. Scenari

emissivi pre-AIA (anno 2010) e post-AIA (anno 2016)" dell'ARPA Regione Puglia. pp.1-44. http://www.gruppoilva.com/items/23/allegati/1/ILVA_SpA_Allegato_Nota_Dir_205_2013.pdf accessed 21/10/2014.

[54] Banini T, Palagiano C. Environment and Health in Italian Cities: the case of Taranto.A. Malik et al. (eds.), *Environmental Deterioration and Human Health*. Springer Science+Business Media: Dordrecht. 2014. DOI 10.1007/978-94-007-7890-0_2.

[55] Mataloni F, Stafoggia M, Alessandrini E, Triassi M, Biggeri A, Forastiere F. Studio di coorte sulla mortalità e morbosità nell'area di Taranto. A color study on mortality and morbidity in the area of Taranto, Southern Italy. *Epidemiol Prev*. 2012; 36(5): 237-252.

[56] Comba P, Conti S, Iavarone I, Marsili G, Musmeci L, Pirastu R. Ambiente e salute a Taranto: evidenze disponibili e indicazioni di sanità pubblica.

[57] Musmeci L, Bianchi F, Carere M, Cori L. Environment and healthin Gela (Sicily): present knowledge and prospects for future studies.

[58] Cernigliaro A, Pollina Addario S, Nicolotti N. Research on the health state of the Sicilian population living in areas at environmental or natural risk: the experience of the Regional Department of Epidemiology.

[59] Pesce Paola, Castaing Marine, Leone Anna, Varvarà Massimo, Dinaro Ylenia, Madeddu Anselmo, Tisano Francesco, Contrino Lia, Tumino Rosario, Sciacca Salvatore. Differenze di incidenza oncologica e mortalità per sede nelle aree metropolitane e nei SIN in Sicilia: dati 2003-2011.http://www.registro-tumori-integrato.it/

[60] Hakkinen PJ. Seveso Disaster, and the Seveso and Seveso II Directives. *Elsevier*. 2005.

[61] Pesatori AC, Consonni D, Bachetti S, Zocchetti C, Bonzini M, Baccarelli A, Bertazzi PA. Short-and long-term morbidity and mortality in the population exposed to dioxin after the "Seveso accident". Industrial Health 2003; 41: 127–138

[62] Corvalán C, Briggs DJ, Zielhuis G; World Health Organization. Decision-making in environmental health: from evidence to action [cited 2014 Sep 12]. accessed 5/10/2014 from: http://apps.who.int/iris/bitstream/10665/42304/2/0419259406_part1.pdf?ua=1.

[63] Ha J. Applying policy and health effects of air pollution in South Korea: focus on ambient air quality standards. Environmental Health and Toxicology. Volume: 29:1-7, Article ID: e2014011, http://dx.doi.org/10.5620/eht.2014.29.e2014011

[64] WHO. Air quality guidelines for particulate matter, ozone, nitrogen dioxide and sulfur dioxide. Global update 2005. http://www.who.int/phe/health_topics/outdoorair/outdoorair_aqg/en/ accessed 11/10/2014.

10

Aldehyde Measurements in Indoor and Outdoor Environments in Central-Southern Spain

Florentina Villanueva, Araceli Tapia,
Inmaculada Colmenar, José Albaladejo,
Beatriz Cabañas and Ernesto Martínez

Additional information is available at the end of the chapter

1. Introduction

Aldehydes are indoor and outdoor chemical pollutants of particular interest due to their potential adverse health effects on humans and to their important role in atmospheric chemistry. In fact, carbonyls are of critical importance since they are the most stable intermediate species in the oxidation of volatile organic compounds. Even in the indoor environments, aldehydes can be released from ozone reactions with unsaturated VOCs [1]. The most important global source of formaldehyde and acetaldehyde in the atmosphere is of secondary origin, and it is the oxidation of natural and anthropogenic hydrocarbons. However, motor vehicle exhausts represent the primary emission source of these pollutants in urban areas [2].

In occupational and residential indoor environments, predominant carbonyls are aldehydes, mainly formaldehyde and acetaldehyde. Formaldehyde, usually the most abundant aldehyde in air, is classified in Group 1 (human carcinogen) by the International Agency for Research on Cancer [3] and acetaldehyde is classified in Group 2B as a possible carcinogenic in humans [4, 5]. Also, benzaldehyde and acrolein are suspected carcinogens and mutagens, as well as other low-molecular-mass aldehydes, which reactivity and possible mutagenicity are similar to those of acetaldehyde [6-8].

Generally speaking, exposure to carbonyls is higher indoors than outdoors. Indoor aldehyde concentrations are usually 2-10 times higher than the outdoor ones [9, 10], indicating the presence of significant indoor sources such as direct emissions and indoor chemical formation although these higher values are also due to the low air exchange rates in the indoor environment [11]. Particularly, formaldehyde is widely used not only in construction (wood processing, furniture, textiles, and carpeting) but also in various industries [12, 13]. It is also a by-product of certain anthropogenic activities (e.g., smoking tobacco, burning automotive (and other) fuels, and residential wood burning) [12]. Formaldehyde is even a component of many consumable household products such as antiseptics, medicines, cosmetics, dish-washing liquids, fabric softeners, shoe-care agents, carpet cleaners, glues and adhesives, lacquers, etc. [14, 15]. Formaldehyde is also employed as food preservative [16, 17]. For these reasons, formaldehyde is generally found in higher concentrations indoors than outdoors.

Because aldehydes may contribute to different diseases and are mainly found in indoor environment, their measurements are of particular interest, especially when it is known that most people in developed countries spend up to 90% of their time indoors [18]. The objective of this study is to measure and analyze the levels of formaldehyde and acetaldehyde and even other carbonyls in indoor and outdoor air in this region from central-southern Spain. The measurements have been carried out in different sampling periods from 2010 to 2014. Very few studies have been carried out in the indoor air in Spain with the aim of determining different air pollutants [19-27] and most of which have been focused on particulate matter [20, 23-26]. In this paper, the levels of carbonyls in different indoor environments in our country are presented for first time. It consists of new field campaigns together with the results obtained from our previous studies [21, 29].

2. Experimental section

2.1. Sampling sites and their characteristics

Ciudad Real is an urban area with around 65,000 inhabitants and is located in the heart of La Mancha region in central-southern Spain (38.59°N, 3.55°W, at approximately 628 m above sea level) in a fairly flat area, 200 km south of Madrid. With a low presence of industry, traffic is the most important source of air pollution in this city [28]. Puertollano (38° 42"N 04° 07"W, at approximately 700 m above sea level) is a very important industrial area with almost 52,000 inhabitants located at 40 km to the southwest of Ciudad Real. The existent industries situated about 5-6 km southeast from the town center include an oil refinery, a petrochemical industry, a nitrogen fertilizer factory, two power plants, and a coal mine, which means an important source of air pollution in this city. Field measurements in the ecological area were conducted on the southwestern border of the park at about 6 km east of the village of Horcajo de los Montes (39.2°N, 04.4°W, 617 m above sea level) (see Fig. 1). Meteorologically, this zone is characterized by very hot and dry summer period with high insulation (Fig. 1).

The first series of field measurements was conducted in Ciudad Real from September 2008 to April 2009 and were focused on different places of the Faculty of Chemistry: two research laboratories, a laboratory used by practical classes, an office, and the bar of the campus. In addition, an outdoor sampling point was selected to measure levels of the aldehydes. Also, indoor and outdoor measurements were carried out in two private homes, one with smokers (home 1) [19]. Indoor measurements were carried out in two points of the homes, the kitchen and the living room.

Figure 1. Situation map of Ciudad Real, Puertollano, and the sampling point in Cabañeros National Park.

In a second series of measurements, aldehydes (formaldehyde, acetaldehyde, acrolein, acetone, propanal, crotonaldehyde, butanal, benzaldehyde, pentanal, tolualdehydes, and hexanal) were measured in six reading rooms of the campus library and also different common rooms of another building of the Faculty of Chemistry. The third series of field measurements was conducted in two classrooms of one school in Ciudad Real and Puertollano and in two homes (living rooms) in Puertollano. All these measurements were carried out in 2012 and 2013. Finally, during August-November 2010 and February-August 2011, carbonyl compounds were measured in the ecological area of Cabañeros National Park [29].

The first series of field measurements was carried out using Analyst® passive samplers, while Radiello® passive samplers were used in the other campaigns. The Analyst® passive sampler was used for assessment only for formaldehyde and acetaldehyde; for this reason and because both aldehydes produce adverse health effect and formaldehyde is the most abundant aldehyde, this study focuses especially on formaldehyde and acetaldehyde, although in some places other carbonyls have also been measured such as the campus library, some places in the Faculty of Chemistry, or Cabañeros National Park.

2.2. Sampling and analytical methods

Radiello® (Fondazione Salvatore Maugeri, Padova, Italy) and Analyst® (Marbaglass, Palombara Sabina, Rome, Italy) passive samplers were used for monitoring carbonyl compounds. The Radiello® passive samplers for carbonyls consist of a stainless steel cartridge filled with

2,4-dinitrophenylhydrazine-coated Florisil® inside a diffusive body, while Analyst® sampler consists of three parts: a polyethylene cylinder, an antiturbulence net (made of silver for outdoor sampling or stainless steel for indoor sampling), and a 2,4-DNPH-coated adsorbent bed (Florisil or silica gel). Analyst® passive samplers were prepared in the laboratory. A detailed description of both passive samplers is given elsewhere [19].

Indoor devices were positioned at a height of 1.5-2.0 m above the floor, in the middle of the room when possible. Outdoor samples were taken simultaneously in the windows or balconies and protected from bad weather conditions by a mountable polypropylene shelter (for Radiello) or stainless steel shelter (for Analyst).

The sampling duration of the Analyst® samplers was 14-20 days, while Radiello® was 7 days. After exposure, the Radiello® cartridges were introduced in their sealed glass tubes, and the Analyst® was cap and stored in the dark and refrigerated until the analysis. Field blanks were transported together with samplers to the sampling point.

The extraction of the hydrazones and the analytical conditions have been described in previous works [19, 29]. Briefly, both Analyst® and Radiello® passive samplers were extracted with 2 ml of acetonitrile (HPLC grade), and the extract was filtered (PTFE 0.45 mm) and analyzed by HPLC (Varian prostar, CA, USA) coupled to a photodiode array detector. For this, 20 μl of the solution obtained after extraction was injected by a sampling loop into a reversed-phase column C-18 (a Varian Microsorb MV 100-5, 25 cm length × 4.6 mm i.d. and a Supelcosil™LC-18 25 cm × 4.6 mm × 5 μm for the Analyst® and for Radiello®, respectively) and detected at a wavelength of 365 nm, according to the literature. The program of the mobile phase was as follows: 0-7 min, 60% acetonitrile (HPLC grade), and 40% water (from a Milli-Q system); 7-20 min, a gradient up to 100% acetonitrile. The flow was 1 ml min^{-1}.

A series of standards (TO11/IP-6A Aldehyde/Ketone-DNPH Mix, Supelco, Bellefonte, USA) containing formaldehyde, acetaldehyde, acrolein, acetone, propanal, crotonaldehyde, butanal, benzaldehyde, isopentanal, pentanal, o-tolualdehyde, m-tolualdehyde, p-tolualdehyde, hexanal, and 2,5-dimethylbenzaldehyde in acetonitrile were used to obtain a five-point calibration curve for each compound in concentration ranges similar to the tested samples (0.2-4 μg ml^{-1}). There were very good linear relationships between concentration and instrumental response for all carbonyls measured ($R^2 > 0.99$).

2.3. Quality assurance

Blank samples, limits of detection (LOD), and reproducibility of the Analyst® and Radiello® passive samplers were assessed for quality assurance. Method detection limits (MDLs) were defined as three times the standard deviation of the blanks. The aldehyde amount in the Analyst® blank samples ranged from 0.06 to 0.11 μg for formaldehyde (estimated air concentration, 0.19-0.35 μg m^{-3}) and from 0.04 to 0.31 μg for acetaldehyde (estimated air concentration, 0.17-1.32 μg m^{-3}). LOD values for blank samples were 0.39 μg for formaldehyde and 0.09 μg for acetaldehyde, corresponding to a concentration of 1.2 and 0.4 μg m^{-3}, respectively, calculated for a period of 14 days. The coefficients of variation in the reproducibility test ($n = 5$) were 5.8% for formaldehyde and 4.5% for acetaldehyde, while for the Radiello® passive samplers ranged from 0.5% (acetaldehyde) to 4.5% (m/p-tolualdehyde) for carbonyl compounds except for acetone, which was 10%. Method detection limits calculated for a sampling period of 7 ranged from 0.01 μg m^{-3} for 2,5-dimethylbenzaldehyde and isopentanal to 0.26 μg m^{-3} for acetaldehyde.

3. Results and discussion

Table 1 shows a summary of the formaldehyde and acetaldehyde concentrations in all sampling sites investigated in Ciudad Real and Puertollano.

3.1. Levels of carbonyls in indoor environments

3.1.1. Faculty of chemistry

The Faculty of Chemistry consists of several buildings, and the laboratories sampled are in a different building to the rest of common areas listed in Table 1. These common areas in the case of Inorganic and Organic Chemistry Departments are halls where offices and laboratories are around or near them.

In the laboratories sampled, the formaldehyde levels varied between 4 and 23.5 μg m^{-3} depending on the laboratory, while those of acetaldehyde varied from 1.3 to 13.2 μg m^{-3}. The class practical laboratory showed the highest formaldehyde and acetaldehyde concentrations, while the lowest concentrations were found in Research Laboratory 1. The difference between Research Laboratory 1 and the other two laboratories (Research Laboratory 2 and class practical lab) was the ventilation, a fact that could explain the differences in the concentrations of the two aldehydes [19]. Although literature information is hardly available for carbonyl compounds measured in laboratories, these measurements are in the range of those determined in research laboratories and class practical laboratories in an academic institute in Fortaleza, Brazil [30], and are lower than those determined in Rio Grande University in Brazil [31] for a research laboratory and class practical laboratories as shown in Table 2 for comparison.

Sampling site	*n*	Formaldehyde			Acetaldehyde			
		Average ± SD	Min	Max	Average ± SD	Min	Max	Reference
Ciudad Real (urban area) *Research Laboratory 1*								[19]
Indoor	11	7.7 ± 3.0	4.0	11.6	5.3 ± 2.1	1.3	7.6	
Outdoor	10	1.5 ± 0.4	0.9	2.3	1.7 ± 0.6	0.8	3.1	
Research Laboratory 2	2	18.8 ± 4.0	16.0	21.6	5.5 ± 0.8	4.9	6.0	[19]
Class practical lab	2	19.6 ± 5.6	15.6	23.5	1.2 ± 1.4	11.2	13.2	[19]

Sampling site	n	Formaldehyde			Acetaldehyde			Reference
		Average ± SD	Min	Max	Average ± SD	Min	Max	
Office	4	11.3 ± 2.0	8.5	13.3	5.7 ± 1.3	5.0	7.7	[19]
Common areas Faculty Chemistry								This study
Inorganic chemistry (2nd floor)	2	9.2 ± 0.6	8.8	9.7	2.0 ± 0.2	1.9	2.1	
Organic chemistry (2nd floor)	2	17.8 ± 0.6	17.4	18.2	2.9 ± 0.2	2.7	3.0	
Library (2nd floor)	2	12.7 ± 0.9	12.1	13.3	2.4 ± 0.07	2.3	2.4	
Reception (1st floor)	2	11.5 ± 0.3	11.2	11.7	2.3 ± 0.006	2.3	2.3	
Hall (1st floor)	2	6.0 ± 0.3	5.7	6.2	1.5 ± 0.03	1.5	1.5	
Coffee area (1st floor)	2	6.1 ± 0.3	5.9	6.2	1.6 ± 0.2	1.4	1.7	
Outdoor	2	1.7 ± 0.4	1.4	2.0	1.8 ± 1.7	0.6	3.0	
Bar	1	13.4	-	-	26.0	-	-	[19]
University library								This study
Ground floor	8	11 ± 1.9	8.4	14.1	5.2 ± 0.9	4.2	6.8	
First floor	8	10.6 ± 1.6	8.6	14.2	5.05 ± 0.8	4.0	5.9	
Second floor	8	11.1 ± 1.4	8.4	15.0	5.1 ± 0.7	4.2	6.1	
Outdoor	4	2.1 ± 0.4	1.6	2.6	2.2 ± 0.5	1.6	2.9	
Home 1								[19]
Living room	1	75	-	-	19.2	-	-	
Kitchen	2	43.7 ± 2.2	42.1	45.2	21.5 ± 4.4	18.4	24.6	
Outdoor	6	2.2 ± 0.5	1.3	2.7	2.4 ± 1.1	0.82	3.55	
Home 2								[19]
Living room	4	46.0 ± 5.6	41.6	53.3	57.4 ± 8.5	43.3	62.2	
Kitchen	2	38.2 ± 3.3	35.8	40.5	69.2 ± 8.2	63.4	75.0	
Outdoor	6	3.1 ± 0.8	1.9	4.2	3.8 ± 2.0	1.2	6.5	
School								This study
Indoor	4	18.9 ± 4.7	14.7	23.1	5.2 ± 1.8	3.6	6.7	
Outdoor	2	1.8 ± 0.8	1.2	2.3	2.1 ± 2.1	0.6	3.6	
Puertollano (industrial area)								
School								This study
Indoor	4	35.4 ± 14.6	21.5	49.2	8.5 ± 5.4	3.7	13	
Outdoor	2	2.0 ± 1.6	0.9	3.1	0.4 ± 0.1	0.3	0.5	

Sampling site	*n*	Formaldehyde			Acetaldehyde			
		Average ± SD	Min	Max	Average ± SD	Min	Max	Reference
Home 1								This study
Indoor	8	24.1 ± 7.5	17.4	37.1	15.2 ± 5.0	9.3	21.4	
Outdoor	8	3.1 ± 0.6	2.4	3.8	2.4 ± 0.9	1.5	3.9	
Home 2								This study
Indoor	8	74.8 ± 27.3	37.9	107.5	23.6 ± 10.7	9.6	36.3	
Outdoor	8	3.2 ± 0.7	2.2	4.3	2.4 ± 0.9	1.2	3.5	
Cabañeros (ecological area)								
Outdoor	46	0.96 ±0.54	b.d.l	2.56	0.79 ± 0.49	0.13	1.89	[29]

Table 1. Formaldehyde and acetaldehyde concentrations (μg m^{-3}) in indoor and outdoor environments

In the office, the levels of formaldehyde and acetaldehyde were 11.3 and 5.7 μg m^{-3}, respectively. Formaldehyde and acetaldehyde concentrations were quite similar to those obtained for the offices with no smokers of a public building in Rome [19] and were lower than the levels found in the offices of Rio Grande University [31]. The values for formaldehyde are also in the range of those measured in offices from different countries from Europe [32].

Regarding common areas, formaldehyde and acetaldehyde concentrations ranged from 5.7 to 18.2 μg m^{-3} and from 1.4 to 3.0 μg m^{-3}, respectively. The lowest levels for both aldehydes were found on the first floor (hall and coffee area). The coffee area is located on the right side of the hall, and it is not separated from this by any wall. The hall has an extension of approximately 350 m^2, and there is not any furniture; there are some tables and chairs and two vending machines in the coffee area only. Because of this, it seems reasonable that the low levels of both aldehydes are found here. The highest levels were registered on the second floor especially in the Organic Chemistry Department, where the levels were similar to those obtained in the research laboratory 2 and the class practical laboratory placed in another building. This fact could be due to the reactions between the variety of organic solvents employed and ozone, but ozone levels are low usually in indoor environments, and they have not been measured in this study. Therefore, care must be taken when assuring this. The library is located beside the Organic Chemistry Department, which could influence the levels of formaldehyde found here. The reception is located on the first floor and presents levels of formaldehyde and acetaldehyde higher than those found on the other two sampling points of the first floor, maybe due to the presence of a big copying machine in the room and the emission of ozone with the consequent reaction with the unsaturated VOCs in the air (such as cleaning products) [1].

Not only formaldehyde and acetaldehyde have been measured in these common areas of the Faculty of Chemistry but also other carbonyls have been identified and quantified and are shown in Table 3. These are acrolein-acetone, crotonaldehyde, benzaldehyde, pentanal, p-tolualdehyde, and hexanal. The most abundant carbonyl was acrolein-acetone (both

Site	Formaldehyde (μg m⁻³)	Acetaldehyde (μg m⁻³)	Activity	Reference
Rio Grande University, Brazil	32.3-41.0	18.3-26.2	Office	[31]
Rome	8.9-9.4	4.2-4.7	Office	[19]
Europe	3-33	-	Offices	[32]
Fortaleza, Brazil	0.32-81.6	1.2-4.43	Research laboratories	[30]
Rio Grande University, Brazil	96.5	79.4	Research laboratory	[31]
Fortaleza, Brazil	3.78-60.75	1.4-3.18	Class practical laboratories	[30]
Rio Grande University, Brazil	56.5-161.5	38.1-91.2	Class practical laboratories	[31]
Modena, Italy	1.7-67.8	-	Libraries	[36]
Strasbourg, France	8.6-94.5	3.7-25.9	Libraries	[37]
Helsinki, Finland	8.1-77.8	3.7-41.5	Living room	[45]
Strasbourg, France	6-93	0-66	Living room	[44]
10 European cities	14.4 ± 4.9-30.7 ± 17.8	9.6 ± 2.3-15.8 ± 5.4	Homes	[43]
Bari, Italy	3.2-49	1.3-23.5	Kitchen	[46]
Paris, France	21.7±1.9	10.1±1.8	Kitchen	[39]
Lisbon, Portugal	6.3-23.8	-	Schools	[49]
11 European cities	5.6-49.7	1.4-29.1	Schools and kindergarten	[32]

Table 2. Formaldehyde and acetaldehyde concentrations measured in different indoor environments in previous studies

appear together in the chromatogram) with a range of concentration between 14.2 and 58.3 $\mu g\ m^{-3}$. The rest of carbonyls present concentrations between 0.2 $\mu g\ m^{-3}$ for benzaldehyde and 4.1 $\mu g\ m^{-3}$ for hexanal (without taking into account formaldehyde and acetaldehyde explained above).

The indoor/outdoor ratio is generally used to infer penetration to indoor environments and indoor sources. An I/O ratio lower and close to one indicates more outdoor sources [33]. In all sites, air concentrations for formaldehyde inside the buildings were between 5 and 13 times higher than outside, which appears to indicate that strong indoor sources exit, clearly determining indoor air concentrations. The same conclusion can be obtained for the pair acrolein-acetone or p-tolualdehyde (I/O = 8-20) and hexanal (I/O = 13-34), but the indoor levels of these two last carbonyls are low in all sampling sites. For the rest of compounds, the main source could be from outdoor even acetaldehyde with a ratio of 2.4-4.4.

Carbonyl	Inorganic hall	Organic hall	Library	Reception	Hall first floor	Coffee area	Outdoor
Formaldehyde	10.1	19.5	13.9	12.6	6.5	6.6	1.5
Acetaldehyde	2.1	3.1	2.5	2.4	1.6	1.7	0.7
Acrolein-acetone	58.3	16.4	14.2	40.1	30.4	22.4	<LOD
Propanal	1.4	0.3	0.2	<LOD	0.9	0.2	<LOD
Butanal	<LOD	<LOD	<LOD	<LOD	<LOD	<LOD	<LOD
Croton-aldehyde	0.5	1.3	0.7	0.9	0.6	0.4	0.4
Benzaldehyde	0.3	0.3	0.2	-	0.2	0.2	<LOD
Pentanal	1.0	0.7	0.8	2.1	0.4	0.4	0.2
p-tolualdehyde	0.8	1.0	0.6	0.5	0.7	0.4	0.05
hexanal	1.6	3.4	2.8	4.1	1.8	1.6	<LOD

Table 3. Carbonyls measured and their concentrations (in $\mu g\ m^{-3}$) in the common areas of Faculty of Chemistry

3.1.2. Bar

In the bar of the campus (about 200 m^2), the concentration of acetaldehyde was much higher than that of formaldehyde. The higher acetaldehyde value could be due to the cigarette smoke and combustion processes such as those carried out in the kitchen of the bar, although acetaldehyde is also present in many foods and alcoholic drinks [3, 5]. These levels of carbonyls are lower than expected. Hodgson et al. [34] estimated that cigarette smoking contributed from 57% to 84% to the total formaldehyde levels in the smoking areas of cafeterias where the formaldehyde concentrations varied between 5 and 42 $\mu g\ m^{-3}$ when 20-100 cigarettes were being smoked. The combustion of 5-10 cigarettes in a room of 30 m^2 increases the formaldehyde concentrations from 240 to 600 $\mu g\ m^{-3}$ [35]. However, formaldehyde levels measured in the bar were low and similar to those found in the teacher's office and in Research Laboratory 1 probably due to mechanical ventilation and the great surface of more than 200 m^2 of the bar.

3.1.3. University campus library

Carbonyl concentrations were also measured in the University Campus Library since paper can emit large amounts of formaldehyde [36]. Six reading rooms of the library and a sampling point outdoor were monitored during four consecutive weeks from February 6 to March 5, 2011. Two passive samplers were placed in two reading rooms on the ground floor and the same on the first and the second floor. The total samples were 28 between indoor and outdoor sampling.

The carbonyl compounds identified in the library were formaldehyde, acetaldehyde, acetone-acrolein, propanal, crotonaldehyde, butanal, benzaldehyde, pentanal, p-tolualdehyde, and hexanal, and their average concentrations were as follows (in $\mu g\ m^{-3}$): 10.9 ± 1.3, 5.1 ± 0.2, 12.4 ± 2.4, 2.5 ± 0.1, 0.4 ± 0.1, 15.4 ± 1.3, 0.55 ± 0.05, 2.9 ± 0.4, 0.8 ± 0.2, and 6.6 ± 1.1, respectively (quoted errors correspond to the standard deviation). Butanal appeared overlapped with an interference (no carbonyl compound) in the samples. Their high levels obtained with respect to the rest of carbonyls together with the UV spectrum indicate that butanal appears overlapped with other carbonyl compound may be 2-butanone. Both compounds appear together in the chromatogram when a standard is introduced under the analytical conditions. Therefore, butanal has not been considered for the discussion. The most abundant carbonyls were acetone-acrolein and formaldehyde. Their concentrations in the individual samples varied from 8.6 to 17.3 μg m^{-3} and from 8.4 to 15 μg m^{-3}, respectively.

Carbonyl	Ground floor	First floor	Second floor	Outdoor
Formaldehyde	11.0 ± 1.9	10.6 ± 1.9	11.1 ± 2.2	2.1 ± 0.4
Acetaldehyde	5.2 ± 0.9	5.1 ± 0.8	5.1 ± 0.7	2.0 ± 0.5
Acrolein-acetone	15.0 ± 1.9	11.1 ± 1.7	11.2 ±1.8	3.8 ± 0.8
Propanal	2.4 ± 0.3	2.6 ± 0.5	2.6 ± 0.4	1.0 ± 0.1
Crotonaldehyde	0.4 ± 0.1	0.5 ± 0.3	0.4 ± 0.3	0.7 ± 0.1
Benzaldehyde	0.5 ± 0.1	0.6 ± 0.1	0.6 ± 0.1	0.3 ± 0.06
Pentanal	3.3 ± 1.0	2.6 ± 0.4	2.7 ± 0.6	1.0 ± 0.1
p-tolualdehyde	0.9 ± 0.5	0.7 ± 0.1	0.8 ± 0.3	0.7 ± 0.3
hexanal	7.7 ± 2.0	5.9 ± 0.9	6.1 ± 0.9	1.7 ± 0.07

Table 4. Levels of carbonyls (in μg m^{-3}) measured in the Campus University Library together with standard deviations

Generally, the concentrations of the carbonyl compounds found in the different reading rooms of the library were low. These values are in the range of those obtained in 20 university libraries in Strasbourg (France) [37], where the levels found ranged from 8.6 to 94.5 for formaldehyde, from 3.7 to 25.9 μg m^{-3} for acetaldehyde, from 2.1 to 58.8 μg m^{-3} for hexanal, from 0.2 to 5.3 μg m^{-3} for benzaldehyde, and from 0.7 to 16.3 μg m^{-3} for propanal. Our measurement values for formaldehyde are also in the range of those reported by Fantuzzi et al. [36] for 16 libraries of the university of Modena (Italy), which ranged from 1.7 to 67.8 μg m^{-3} with an average value of 32.7 ± 23.9 μg m^{-3} (see Table 2). However, the average concentrations for formaldehyde calculated in this study are lower than the studies mentioned above. Formaldehyde, acetaldehyde, propanal, benzaldehyde, and hexanal have usually been detected in libraries [37].

Table 4 shows the concentrations of all carbonyls measured on the different floors of the library. There are not differences between the concentrations measured of the carbonyl compounds in the different reading rooms of the different floors of the library. The indoor/outdoor ratios vary from 0.65 crotonaldehyde, which indicates mainly outdoor sources to 5.1 for formaldehyde indicating mainly indoor sources. Other I/O ratios that could suggest indoor sources but are not strong are 3.3 for acetone and 3.8 for hexanal.

3.1.4. Private houses

Formaldehyde and acetaldehyde measurements were performed simultaneously in the living room and outdoor in Ciudad Real and Puertollano; aldehydes in the kitchens were also measured only in Ciudad Real (see Table 1). Homes were ventilated as usual by opening the windows during the sampling. Two samples were collected at the same time in Ciudad Real, while the samples were collected during 8 consecutive weeks in Puertollano. All houses were located in an area of moderated traffic. Approximate home ages are between 1 and 8 years old in Ciudad Real and Puertollano; only home 2 in Ciudad Real is more than 10 years. This information is important since building materials can emit formaldehyde, but these emissions generally decrease with the age of the house. Smoking people are present in one house of each town.

Formaldehyde indoor air concentration was in average 24.5 and 15.3 times higher than the concentration outdoor in Ciudad Real and Puertollano, respectively, while acetaldehyde was 11.5 and 7.9 times higher in Ciudad Real and Puertollano, respectively. These ratios are similar to those found by Geiss et al. [32] for a new manufacturer house (20.2 for formaldehyde and 9.2 for acetaldehyde); however, the homes in Ciudad Real are more than five years old and between 1 and 8 in Puertollano. This fact indicates that strong indoor sources existed, clearly determining indoor air concentrations and denoted that indoor sources were dominant for these compounds. Indoor materials such as consumer products, furniture, and decorations are important indoor sources [38-41]. Carbonyls such as acetaldehyde are generated from smoking [42]. A smoker person lives in home 2 (Ciudad Real) and presented the highest acetaldehyde levels in both kitchen and living room. Home 1 (Puertollano) is also the home of one smoker, but the frequency of smoking is low; in this case, the levels of acetaldehyde are even lower than the levels obtained in the homes of nonsmokers. Except for home 2 from Ciudad Real, formaldehyde levels were higher than the acetaldehyde ones. Bruinen de Bruin et al. [43] measured formaldehyde and acetaldehyde concentrations in residential environments in ten European cities, demonstrating mean levels in the range of 14.4 ± 4.9 μg m^{-3} (Dublin) to 30.7 ± 17.8 μg m^{-3} (Arnhem) for formaldehyde and from 9.6 ± 2.3 to 15.8 ± 5.4 μg m^{-3} (Budapest) for acetaldehyde, thus confirming low concentrations of aldehydes in European homes. Another recent study measured formaldehyde and acetaldehyde inside the homes of Strasbourg (France) [44]. The indoor levels were in the range of 6 to 93 μg m^{-3} for formaldehyde and from 0 to 66 μg m^{-3} for acetaldehyde. Therefore, the mean concentration of formaldehyde in the indoor air in the present work is in the range of those found in other European cities such as Strasbourg or Helsinki (8.1-77.8 μg m^{-3}) [45] (see Table 2). Indoor acetaldehyde had also a mean concentration similar to that obtained in other European cities (see Table 2), except for home 2 in Ciudad Real wherein concentration is the highest with an average of 57.4 μg m^{-3}. This concentration is in agreement with some living rooms in Strasbourg [44], where the range of concentrations was from 0 to 66 μg m^{-3} for acetaldehyde.

In the case of concentrations of aldehydes measured in the kitchens of Ciudad Real, formaldehyde presents practically the same level for both homes. Our mean value is approximately 40% higher than that reported in previous studies [39] and is in the range of that reported in reference [46] for a study performed in Bari (Italy). For acetaldehyde, the mean concentration registered in the kitchen home 2 is higher than those reported in other European cities.

According to the World Health Organization, the lowest formaldehyde concentration that has been associated with nose and throat irritation after short-term exposure is 100 μg m^{-3} [47]. Therefore, it represents the recommended maximal value. In our study, there is no such place with formaldehyde values exceeding this level. Despite that, it is important to note that the sampling period was 14 days for the Analyst® passive sampler (homes in Ciudad Real) and 7 days for Radiello® (homes in Puertollano). Thus, our results express an average concentration over a long period and do not provide information about exposure peaks, which must be higher than the average concentration measured. However, other recommended values exist, which were exceeded in our study. For example, the indoor formaldehyde concentrations in the homes monitored exceeded the guideline value of 30 μg m^{-3} proposed for the prevention of irritant effects [48] referred to a period of 30 min.

Hence, it is necessary to carry out an exhaustive research with more homes monitored in order to assess the health risk. In addition, taking into account the low outdoor pollution levels observed, an adequate airing of the rooms could reduce the indoor pollution by formaldehyde and acetaldehyde. Nevertheless, this study only presents preliminary results of the private homes in Ciudad Real and Puertollano due to the low number of samples. Measurements in Ciudad Real were performed only to check the viability of the Analyst® passive sampler in the indoor environment [19]. Regarding homes in Puertollano, this study presents the preliminary results of a further research. There are not clear differences between the levels of formaldehyde and acetaldehyde registered in the urban and industrial area.

3.1.5. Schools

Two schools were sampling for the present study, one in Ciudad Real and one in Puertollano located in the same area as the homes sampled. Samples were simultaneously collected in two classrooms from each school and outside the building. Every classroom has an area of about 50 m^2, and there are between 20 and 27 children. The school in Puertollano is more than 20 years old, and the classrooms have not been renovated recently, while in Ciudad Real, the school is also old but the classrooms were renovated 7 years ago. The ventilation in both schools is natural by opening the windows during 15 min per day.

The concentration of formaldehyde and acetaldehyde was slightly higher for the industrial area 35.4 ± 14.6 and 8.5 ± 5.4 μg m^{-3}, respectively, versus 24.3 ± 4.7 and 6.2 ± 1.8 μg m^{-3}, respectively, for the urban area. The concentrations of formaldehyde were more abundant than the acetaldehyde ones. In both schools, the air concentrations for both aldehydes inside the buildings were higher than outside. The indoor/outdoor ratios for formaldehyde and acetaldehyde in Ciudad Real were 13.5 and 2.9, respectively, and were higher for both aldehyde in the school from the industrial area 17.7 and 21 for formaldehyde and acetaldehyde, respectively. Except acetaldehyde in the school from Ciudad Real, formaldehyde and acetaldehyde are mainly indoor pollutants derived from indoor sources.

Our data for formaldehyde are similar or a bit higher than the levels registered in fourteen schools in Lisbon (Portugal) [49]. The levels of both aldehydes are in agreement with the concentrations registered in eleven European cities [32] (see Table 2).

The levels of formaldehyde exceeded the guideline value of 30 μg m^{-3} proposed for the prevention of irritant effects [48] in the industrial area. High levels of formaldehyde are likely associated with the age of building and renovating activities of old buildings. However, both schools are more than 10 years old, and they did not have recent renovation therefore, these levels are due to other sources.

3.2. Levels of carbonyls in outdoor environments

In all sampling sites, formaldehyde and acetaldehyde were present in both indoor and outdoor air. Generally, in all sites, air concentrations for these aldehydes inside the buildings were higher than outside. Mean values for formaldehyde in Ciudad Real varied between 1.5 and 3.1 μg m^{-3}, while acetaldehyde varied between 1.7 and 3.8 μg m^{-3} for the sampling period of December-February. Outdoor acetaldehyde concentrations were similar to those of formaldehyde. In the case of the industrial area, formaldehyde varied in the range of 2 (in March) and 3.2 μg m^{-3} (May-June) and acetaldehyde varied between 0.4 (in March) and 2.4 μg m^{-3} (May-June), being formaldehyde levels slightly higher than the acetaldehyde ones. Similar values were found in samplings conducted in rural and semirural areas [50-52].

Formaldehyde/acetaldehyde ratio has been proposed as an indicator of the biogenic source of formaldehyde [53] and can vary between 1 (urban area) and 10 (deciduous forest). This ratio for Ciudad Real and Puertollano is between 0.9 and 1.3 reflecting typical values for urban air. The formaldehyde/acetaldehyde ratio for the sample taken in the school in Puertollano gave a value of 5, which could imply a biogenic contribution in the early spring.

A study about levels of carbonyls was carried out in the ecological area of Cabañeros National Park [29]. Twelve compounds were identified and quantified: formaldehyde, acetaldehyde, acetone-acrolein, propanal, crotonaldehyde, butanal, benzaldehyde, isopentanal, pentanal, o-tolualdehyde, m/p-tolualdehyde, and hexanal (the sum of m/p-tolualdehyde and acetone-acrolein was reported because they could not be well separated by the analytical method). The most abundant carbonyls were hexanal, acetone-acrolein, formaldehyde, and acetaldehyde.

Because of the study covered all seasons, August 2010 was the most carbonyl-polluted month followed by July 2011 and September 2010. October and November 2010 were the months with lower concentration of carbonyl compounds showing an apparent seasonal variation with maximum values observed in summer months. The concentrations of the carbonyls were as follows: acetone-acrolein mixing ratios ranged from 0.35 in February to 4.52 μg m^{-3} in June 2011, with an average of 1.78 μg m^{-3}, while ambient levels of hexanal varied from 0.67 μg m^{-3} in October to 1.72 μg m^{-3} in April; the average concentration during the sampling period was 1.06 μg m^{-3}. Formaldehyde was in the range of values below detection limit and 2.56 μg m^{-3} in October and June, respectively, with an average concentration of 0.96 μg m^{-3} and acetaldehyde varied from levels below detection limits in February, March, and April to 1.89 μg m^{-3} in June. The concentrations of other carbonyls ranged from non-detected to 1.18 μg m^{-3}.

The seasonal cycles obtained for formaldehyde, acetaldehyde, and acetone-acrolein with respect to the other carbonyls suggest different formation mechanism and sinks compared to the others. This could be due to photochemical processes (oxidation of biogenic and even anthropogenic hydrocarbons that arrive to the area under determined meteorological conditions) and also the direct emission from vegetation. For example, two Mediterranean tree species such as *Quercus ilex* (holm oak), one of the most abundant oak in the study area, and *Pinus pea* (Italian stone pine) emit formaldehyde and acetaldehyde [54, 55].

The formaldehyde and acetaldehyde concentrations measured at Cabañeros National Park are in the same range or lower than the levels reported in other forests or rural areas (Germany [56] and Brazil [57]). Our data are similar to those reported in the small village of Covelo in Portugal, considered as a rural/forest site [58].

On the other hand, the levels found in the ecological area during the months of February and March for formaldehyde (0.51 and 0.59 μg m^{-3}) and acetaldehyde (0.33 and 0.29 μg m^{-3}) are lower than those registered in the urban (outdoor samples collected at the same time as common areas of Faculty of Chemistry or University Library) and industrial area (outdoor samples collected in the school) during the same period. Also, the levels of both aldehydes during the months of May and June were lower (1.15-1.37 μg m^{-3} for formaldehyde and 0.70-1.09 μg m^{-3} for acetaldehyde) than the levels registered in the industrial area during the same months in the outdoor samples collected in the homes.

3.3. Health risk evaluation

Formaldehyde is classified in Group 1 by IARC. It was based on inhalation causing squamous cell carcinoma in rats and nasopharyngeal cancer in humans [3]. Recently, the classification has been expanded with formaldehyde causing leukemia and limited evidence of sinonasal cancer in humans [59]. The lowest concentration reported to cause sensory irritation of the eyes in humans is 0.38 mg m^{-3} for 4 h. Increases in eye blink frequency and conjunctival redness appear at 0.6 mg m^{-3}, which is considered equal to the no-observed-adverse-effect level (NOAEL). There is no indication of accumulation of effects over time with prolonged exposure, and there is no evidence indicating an increased sensitivity to sensory irritation to formaldehyde among people often regarded as susceptible (asthmatics, children, and older people) [60].

The formaldehyde exposure-response relationship is highly nonlinear and biphasic, supporting a NOAEL that allows setting a guideline value [60]. This means that it cannot be assumed a low-dose linear relationship for the carcinogenic effects, and therefore, the calculation of the incremental lifetime risk of cancer for formaldehyde, assuming a low-dose linear relationship, would give a value highly overestimated and it would be meaningless.

As commented above, WHO established a guideline value for formaldehyde of 100 μg m^{-3} that should not be exceeded for any 30-min period of the day [47]. This short-term guideline will also prevent effects on lung function as well as long-term health effects, including nasopharyngeal cancer and myeloid leukemia [60].

We can compare our data with the short-term guideline value of 100 μg m^{-3}. In our study, the maximum values registered in the different environments were below 100 μg m^{-3}, except for home 2 in the industrial area that registered a maximum value of 108 μg m^{-3} during one of the 8 weeks sampled. The average concentrations for formaldehyde found in this study ranged from 6 to 75 μg m^{-3} in indoor environments, while outdoor concentrations were considerably lower. These values are below the guideline value considered preventive of carcinogenic effects; however, our results represent an average value during 1 or 2 weeks of exposition, and this guideline value is referred to a period of 30 min. Nevertheless, as long as formaldehyde concentrations are below 100 μg m^{-3}, there should be no chance of developing cancer.

On the other hand, the most important way to control the indoor formaldehyde concentration is the air exchange rate and the use of low-emitting materials and products. Environmental tobacco smoke and ozone-initiated reactions of alkene compounds may contribute to temporary peak levels [60]

4. Conclusion

This paper presents an overview of the concentrations of priority aldehydes observed in indoor and outdoor air in an urban and an industrial area of central-southern Spain. Very few studies have been carried out in the indoor air of Spain. Therefore, we have tried to gather the data previously published about formaldehyde and acetaldehyde in this region and other new data to obtain a general view of the levels in different outdoor and indoor environments.

From the results, we can conclude that the highest concentrations of formaldehyde and acetaldehyde are found inside the homes and schools. The levels of formaldehyde in laboratories or in rooms near laboratories, where many kinds of solvents are used, are much lower. Overall indoor and outdoor concentrations of formaldehyde and acetaldehyde in all sites sampled were below the threshold limit of 100 μg m^{-3} proposed by WHO [47] associated with nose and throat irritation and also to prevent all types of cancer. Therefore, it represents the recommended maximal value. In our study, there is no such place with formaldehyde values exceeding this level. Despite that, it is important to note that the sampling period was 14 and 7 days for Analyst® and Radiello®, respectively. Thus our results express an average concentration over a long period and do not provide information about exposure peaks, which must be much higher than the average concentration measured. On the other hand, the indoor formaldehyde concentrations in the homes and one school monitored exceeded the recommended guideline value of 30 μg m^{-3} proposed by Kotzias et al. [48] for the prevention of irritant effects. Also this value is referred to a period of 30 min.

Therefore, although the indoor sources are dominant for all sites sampled, the most important to be studied are at homes and schools. This is only a preliminary study about the levels of formaldehyde and acetaldehyde in homes and schools due to the low number of samples, and an exhaustive research is necessary in order to better characterize the chemical composition of the air that people breathe daily and to assess the health risk.

Author details

Florentina Villanueva[1,2*], Araceli Tapia[3], Inmaculada Colmenar[3], José Albaladejo[3], Beatriz Cabañas[1] and Ernesto Martínez[1]

*Address all correspondence to: florentina.vgarcia@uclm.es

1 Atmospheric Pollution Laboratory, Research Institute for Combustion and Atmospheric Pollution, University of Castilla La Mancha, Ciudad Real, Spain

2 Castilla La Mancha Science and Technology Park, Spain

3 Physical Chemistry Department, Faculty of Chemical Science and Technologies, University of Castilla La Mancha, Ciudad Real, Spain

References

[1] Zhang J, Wilson WE, Lloy PJ. Indoor air chemistry: formation of organic acids and aldehydes. Environmental Science and Technology 1994;28 1975-1982.

[2] Granby K, Carsten SC, Lohse C. Urban and semi-rural observations of carboxylic acids and carbonyls. Atmospheric Environment 1997;31 1403-1415.

[3] IARC. IARC monographs on the evaluation of carcinogenic risks to human. Formaldehyde, 2-butoxyethanol and 1-tert-butoxypropan-2-ol. IARC, Lyon, France, 2006;88 39-325.

[4] IARC. IARC monographs on the evaluation of carcinogenic risks to human. Re-evaluation of some organic chemicals, hydrazine and hydrogen peroxide: acetaldehyde. IARC, Lyon, France, 1999;71.

[5] WHO, Acetaldehyde. Environmental Health Criteria, No. 167, World Health Organization, Geneva, 1995.

[6] Liu W, Zhang J, Korn LR, Zhang L, Weisel CP, Turpin B, Morandi M, Stock T, Colome S. Predicting personal exposure to airborne carbonyls using residential measurements and time/activity data. Atmospheric Environment 2007;41 5280-5288.

[7] Tsai SW, Chang TA. Time-weighted average sampling of airborne n-valeraldehyde by a solid-phase microextraction device. Journal of Chromatography A 2002;954 191-198.

[8] US EPA 2003. Integrated Risk Information System. Washington, DC. http://cfpub.epa.gov/ncea/iris/

[9] Zhang F, He Q, Lioy PJ. Characteristics of aldehydes: concentrations, sources, and exposures for indoor and outdoor residential microenvironments. Environmental Science and Technology 1994;28 146-152.

[10] Baez A, Padilla H, Garcia R, Torres MDC, Rosas I, Belmont R. Carbonyl levels in indoor and outdoor air in Mexico City and Xalapa, Mexico. Science of the Total Environment 2003;302 211-226.

[11] Salthammer T, Fuhrman F, Kaufhold S, Meyer B, Schwarz A. Effect of climate parameters on formaldehyde concentrations in indoor air. Indoor Air, 1995;5, 120.

[12] ATSDR. Agency for Toxic Substances and Disease Registry. Toxicological Profile for Formaldehyde. Agency for Toxic Substances and Disease Registry, US Department of Health and Human Services, 1999.

[13] Bolt HM, Degen GH, Hengstler JG. The carcinogenicity debate on formaldehyde: how to derive safe exposure limits? Archives of Toxicology 2010;84(6) 421-422.

[14] Bracamonte BG, Ortiz FJ, Diez LI. Occupational allergic contact dermatitis due to formaldehyde and textile finish resins. Contact Dermatitis 1995;33 139-140.

[15] CARB. California Air Resources Board. Determination of Formaldehyde and Toluene Diisocyanate Emissions from Indoor Residential Sources. Research Division, California Air Resources Board, Sacramento, CA, 1996.

[16] Cai Y, Lin S, Jin Y, Chen H, Zheng J, Chen B. Investigation of formaldehyde level in marketed marine products. Strait Journal of Preventive Medicine 2000;6(5) 32-39.

[17] Naya M, Nakanish J. Risk assessment of formaldehyde for the general population in Japan. Regulatory Toxicology and Pharmacology 2005;43(3) 232-248.

[18] WHO. Environment and Health Risks: A Review of the Influence and Effects of Social Inequalities. World Health Organization, Copenhagen, Denmark, 2010.

[19] Villanueva F, Colmenar I, Mabilia R, Scipioni C, Cabañas B. Field evaluation of the Analyst® passive sampler for the determination of formaldehyde and acetaldehyde in indoor and outdoor ambient air. Analytical Methods 2013;5 516-524.

[20] Rivas I, Viana M, Moreno T, Pandolfi M, Amato F, Reche C, Bouso L, Alvarez-Pedrerol M, Alastuey A, Sunyer J, Querol X. Child exposure to indoor and outdoor air pollutants in schools in Barcelona, Spain. Environment International 2014;69 200-212.

[21] Nadala M, Inzab I, Schuhmachera M, Figuerasb MJ, Domingo JL. Health risks of the occupational exposure to microbiological and chemical pollutants in a municipal waste organic fraction treatment plant. International Journal of Hygiene and Environmental Health 2009;212 661-669.

[22] de Blas M, Navazo M, Alonso L, Durana N, Gomez MC, Iza J. Simultaneous indoor and outdoor on-line hourly monitoring of atmospheric volatileorganic compounds in

an urban building. The role of inside and outside sources. Science of the Total Environment 2012;426 327-335.

[23] Moreno T, Rivas I, Bouso L, Viana M, Jones T, Alvarez-Pedrerol M, Alastuey A, Sunyer J, Querol X. Variations in school playground and classroom atmospheric particulate chemistry. Atmospheric Environment 2014;91 162-171.

[24] Viana M, Rivas I, Querol X, Alastuey A, Sunyer J, Álvarez-Pedrero M, Bouso L, Sioutas C. Indoor/outdoor relationships and mass closure of quasi-ultrafine, accumulation and coarse particles in Barcelona schools. Atmospheric Chemistry and Physics 2014;14 4459-4472.

[25] Minguillón MC, Schembari A, Triguero-Mas M, de Nazelle M, Dadvand P, Figueras F, Salvado JA, Grimalt JO, Nieuwenhuijsen M, Querol X. Source apportionment of indoor, outdoor and personal PM2.5 exposure of pregnant women in Barcelona, Spain. Atmospheric Environment 2012;58 426-436.

[26] Horemansa B, Cardell C, Bencs L, Kontozova-Deutsch V, De Wael K, Van Grieken R. Evaluation of airborne particles at the Alhambra monument in Granada, Spain. Microchemical Journal 2011;99 429-438.

[27] Parra MA, Elustondo D, Bermejo R, Santamaría JM. Exposure to volatile organic compounds (VOC) in public buses of Pamplona, Northern Spain. Science of the Total Environment 2008;404 18-25.

[28] Villanueva F, Notario A, Albaladejo J, Millán MC, Mabilia R. Ambient air quality in an urban area (Ciudad Real) in central-southern Spain. Fresenius Environmental Bulletin 2010;19b 2064-2070.

[29] Villanueva F, Tapia A, Notario A, Albaladejo J, Martínez E. Ambient levels and temporal trends of VOCs, including carbonyl compounds, and ozone at Cabañeros National Park border, Spain. Atmospheric Environment 2014;85 256-265.

[30] Cavalcante RM, Campelo CS, Barbosa MJ, Silveirab ER, Carvalho TV, Nascimento RF. Atmospheric Environment 2006;40 5701-5711.

[31] Cavalcante RM, Seyffert B, Montes D'Oca M, Nascimento RF, Campelo CS, Pinto IS, Anjos FB, Costa AHR. Exposure assessment for formaldehyde and acetaldehyde in the workplace. Indoor Built Environment 2005;14 165-172.

[32] Geiss O, Giannopoulos G, Tirendi S, Barrero-Moreno J, Larsen BR, Kotzias D. The AIRMEX study-VOC measurement in public buildings and schools/kindergartens in eleven European cities: statistical analysis of the data. Atmospheric Environment 2011;45 3676-3684.

[33] Edwards RD, Jurvelin, J, Koistinen K, Saarela K, Jantunen M. VOC source identification from personal and residential indoor, outdoor, and workplace microenvironment samples in EXPOLIS-Helsinki, Finland. Atmospheric Environment 2001;35 4829-4841.

[34] Hodgson, AT, Beal D, McIlvaine JER. Sources of formaldehyde, other aldehydes and terpenes in a new manufactured house. Indoor Air 2002;12 235-242.

[35] Grimaldi F, Botti P, Bouthiba M, Gouezo F, Viala A. Study of indoor air pollution by carbonyl compounds. Pollution Atmospherique 1996;38(149) 58-67.

[36] Fantuzzi G, Aggazzotti G, Righi E, Cavazzuti LGP, Franceschelli A. Indoor air quality in the university libraries of Modena (Italy). Science of the Total Environment 1996;19349-56.

[37] Allou L, Marchand C, Mirabel Ph, Le Calve S. Aldehydes and BTEX measurements and exposures in university libraries in Strasbourg (France). Indoor and Built Environment 2008;17(2) 138-145.

[38] Baumann MGD, Lorenz LF, Batterman SA, Zhang GZ. Aldehyde emissions for particleboard and medium density fiber board products. Forest Product Journal 2000;50 75-82.

[39] Clarisse B, Laurent AM, Seta N, Moullec LY, Hasnaoui EA, Momas I. Indoor aldehydes: measurement of contamination levels and identification of their determinants in Paris dwellings. Environmental Research 2003;92 245-253.

[40] Fjällström P, Andersson B, Nilsson C. Drying of linseed oil paints: the effects of substrate on the emission of aldehydes. Indoor Air 2003;13 277-282.

[41] Knudsen HN, Kjaer UD, Nielsen PA, Wolkoff P. Sensory and chemical characterization of VOC emissions from building products: impact of concentration and air velocity. Atmospheric Environment 1999;33 1217-1230.

[42] Baek SK, Jenkins RA. Characterization of trace organic compounds associated with aged and diluted sidestream tobacco smoke in a controlled atmosphere—volatile organic compounds and polycyclic aromatic hydrocarbons. Atmospheric Environment 2004;38 6583-6599.

[43] Bruinen de Bruin Y, Koistinen K, Kephalopoulos S, Geiss S, Tirendi S, Kotzias D. Characterisation of urban inhalation exposures to benzene, formaldehyde and acetaldehyde in the European Union. Environmental Science and Pollution Research 2008;15 417-430.

[44] Marchand C, Le Calvé S, Mirabel Ph, Glasser N, Casset A, Schneider N, de Blay F. Concentrations and determinants of graseous aldehydes in 162 homes in Strasbourg (France). Atmospheric Environment 2008;42 505-516.

[45] Juverlin J, Vartiainen M, Jantunen M. Personal exposure levels and microenvironmental concentrations of formaldehyde and acetaldehyde in the Helsinki Metropolitan Area, Finland. Journal of the Air and Waste Management Association 2001;51 17-24.

[46] Lovreglio P, Carrus A, Iavicoli S, Drago I, Persechino B, Soleo L. Indoor formaldehyde and acetaldehyde levels in the province of Bari, South Italy, and estimated health risk. Journal of Environmental Monitoring 2009;11 955-961.

[47] WHO. Air Quality Guidelines for Europe. World Health Organization, Regional Office for Europe, Copenhagen, Denmark, 2nd ed., 2001.

[48] Kotzias D, Koistinen K, Kephalopoulos S, Schlitt C, Carrer P, Maroni M, Jantunen M, Cochet C, Kirchner S, Lindvall T, McLaughlinc J, Molhavec L, Fernandes E, Seifert B. The INDEX Project. Critical Appraisal of the Setting and Implementation of Indoor Exposure Limits in the EU, European Commission, Directorate General, Joint Research Centre, EUR 21590 EN, 2005

[49] Pegas PN, Alves CA, Evtyugina MG, Nunes T, Cerqueira M, Franchi M, Pio CA, Almeida SM, Cabo Verde S, Freitas MC. Seasonal evaluation of outdoor/indoor air quality in primary schools in Lisbon. Journal of Environmental Monitoring 2011;13 657-667.

[50] Christensen CS, Skov H, Nielsen T, Lohse C. Temporal variation of carbonyl compound concentrations at a semi-rural site in Denmark. Atmospheric Environment 2000;34 287-296.

[51] Gaffney JS, Marley NA, Martin RS, Dixon RW, Reyes LG, Popp CJ. Environmental Science and Technology 1997;31 3053-3061.

[52] Possanzini M, Tagliacozzo G, Cecinato A. Ambient levels and sources of lower carbonyls at Montelibretti, Rome (Italy). Water, Air and Soil Pollution 2007;183 447-454.

[53] Shepson PB, Hastie DR, Schiff HI, Polizzi M, Bottenheim JW, AnlaufK, Mackay GI, Karecki DR. Atmospheric concentrations and temporal variations of C_1-C_3 carbonyl compounds at two rural sites in central Ontario. Atmospheric Environment 1991;25A 2011-2015.

[54] Schäfer L, Gabriel R, Müller H, Wolf A, Kesselmeier J. Emission of short chained organic acids and aldehydes in relation to physiological activities and apoplastic ion concentration in Mediterranean tree species during the B.E.M.A. field campaign in May 1994. In: Biogenic Emissions in the Mediterranean Area, BEMA-Project, Report on the 1st BEMA measuring campaign at Castelporziano, Rome (Italy), May 1994, EUR 16293 EN, Brussels, 1995;233-247.

[55] Kesselmeier J, Bode K, Hofmann U, Müller H, Schäfer L, Wolf A, Ciccioli P, Brancaleoni E, Cecinato A, Frattoni M, Foster P, Ferrari C, Jacob V, Fugit JL, Dutaur L, Simon V, Torres L. The BEMA-Project: emission of short chained organic acids, aldehydes and monoterpenes from *Quercus ilex* L. and *Pinuspinea* L. in relation to physiological activities, carbon budget and emission algorithms. Atmospheric Environment 1997;31 119-134.

[56] Müller K, Haferkorn S, Grabmer W, Wisthaler A, Hansel A, Kreuzwieser J. Biogenic carbonyl compounds within and above a coniferous forest in Germany. Atmospheric Environment 2006;40 S81-S91.

[57] Custódio D, Guimaraes CS, Varandas L, Arbilla G. Pattern of volatile aldehydes and aromatic hydrocarbons in the largest urban rainforest in the Americas. Chemosphere 2010;79 1064-1069.

[58] Evtyugina MG, Nunes T, Pio C, Costa CS. Photochemical pollution under sea breeze conditions, during summer at the Portuguese West Coast. Atmospheric Environment 2006;40 6277-6293.

[59] IARC, IARC monographs on the evaluation of carcinogenic risks to human. Formaldehyde. IARC, Lyon 100F:1-36. Monographs.iarc.fr/ENG/Monographs/vol100F/mono100F-29.pdf. Accessed June 17, 2015.

[60] WHO. Guidelines for Indoor Air Quality: Selected Pollutants. World Health Organization, Regional Office for Europe, Copenhagen, Denmark, 2010.

Permissions

All chapters in this book were first published in CAQI, by InTech Open; hereby published with permission under the Creative Commons Attribution License or equivalent. Every chapter published in this book has been scrutinized by our experts. Their significance has been extensively debated. The topics covered herein carry significant findings which will fuel the growth of the discipline. They may even be implemented as practical applications or may be referred to as a beginning point for another development.

The contributors of this book come from diverse backgrounds, making this book a truly international effort. This book will bring forth new frontiers with its revolutionizing research information and detailed analysis of the nascent developments around the world.

We would like to thank all the contributing authors for lending their expertise to make the book truly unique. They have played a crucial role in the development of this book. Without their invaluable contributions this book wouldn't have been possible. They have made vital efforts to compile up to date information on the varied aspects of this subject to make this book a valuable addition to the collection of many professionals and students.

This book was conceptualized with the vision of imparting up-to-date information and advanced data in this field. To ensure the same, a matchless editorial board was set up. Every individual on the board went through rigorous rounds of assessment to prove their worth. After which they invested a large part of their time researching and compiling the most relevant data for our readers.

The editorial board has been involved in producing this book since its inception. They have spent rigorous hours researching and exploring the diverse topics which have resulted in the successful publishing of this book. They have passed on their knowledge of decades through this book. To expedite this challenging task, the publisher supported the team at every step. A small team of assistant editors was also appointed to further simplify the editing procedure and attain best results for the readers.

Apart from the editorial board, the designing team has also invested a significant amount of their time in understanding the subject and creating the most relevant covers. They scrutinized every image to scout for the most suitable representation of the subject and create an appropriate cover for the book.

The publishing team has been an ardent support to the editorial, designing and production team. Their endless efforts to recruit the best for this project, has resulted in the accomplishment of this book. They are a veteran in the field of academics and their pool of knowledge is as vast as their experience in printing. Their expertise and guidance has proved useful at every step. Their uncompromising quality standards have made this book an exceptional effort. Their encouragement from time to time has been an inspiration for everyone.

The publisher and the editorial board hope that this book will prove to be a valuable piece of knowledge for researchers, students, practitioners and scholars across the globe.

List of Contributors

Veronica Distefano, Sandra De Iaco, Monica Palma, Donato Posa and Alessandra Spennato
University of Salento, Department of Management, Economics, Mathematics and Statistics, Italy

Oyewale Mayowa Morakinyo
Department of Environmental Health Sciences, Faculty of Public Health, University of Ibadan, Nigeria

Godson Rowland Ana
Department of Environmental Health Sciences, Faculty of Public Health, University of Ibadan, Nigeria
Climate and Society Programme Centre for Sustainable Development, University of Ibadan,Nigeria

Gregory Adekunle Fakunle
Department of Environmental Health Sciences, Faculty of Public Health, University of Ibadan, Nigeria
Department of Medicine, University College Hospital, Ibadan, Nigeria

Davidson Martins Moreira
Center for Integrated Manufacturing and Technology (CIMATEC), Salvador, Brazil

Tiziano Tirabassi
Institute ISAC of CNR, Bologna, Italy

Taciana Albuquerque
Universidade Federal de Minas Gerais, Belo Horizonte, Brazil

Wenqing Liu, Zhenyi Chen, Pinhua Xie, Jianguo Liu and Tianshu Zhang
State Environmental Protection Key Laboratory of Optical Monitoring Technology, Ministry of Environmental Protection of the People's Republic of China, Anhui Institute of Optics and Fine Mechanics, Chinese Academy of Sciences, Hefei, Anhui, P.R. China

Anca Maria Moldoveanu
Carol Davila University of Medicine and Pharmacy, Bucharest, Romania
National Institute of Public Health, Bucharest, Romania

Farideh Golbabaei and Monireh Khadem
Department of Occupational Health Engineering, School of Public Health, Tehran University of Medical Sciences, Tehran, Iran

Chang-Jin Ma
Department of Environmental Science, Fukuoka Women's University, Fukuoka, Japan

Mariko Yamamoto
Department of Environmental Science, Fukuoka Women's University, Fukuoka, Japan
Fukuoka Prefectural Munakata High school, Munakata, Japan

Danni Guo
Statistical Ecology Division, Biodiversity Research, Assessment and Monitoring, South African National Biodiversity Institute, Cape Town, South Africa

M. Ferrante, M. Fiore, C. Copat, S. Morina, C. Ledda, C. Mauceri and G. Oliveri Conti
Environmental and Food Hygiene Laboratories (LIAA), Department of Medical Science, Surgical and Advanced Technologies "GF Ingrassia", Hygiene and Public Health, University of Catania, Italy

Beatriz Cabañas and Ernesto Martínez
Atmospheric Pollution Laboratory, Research Institute for Combustion and Atmospheric Pollution, University of Castilla La Mancha, Ciudad Real, Spain

Florentina Villanueva
Atmospheric Pollution Laboratory, Research Institute for Combustion and Atmospheric Pollution, University of Castilla La Mancha, Ciudad Real, Spain
Castilla La Mancha Science and Technology Park, Spain

Araceli Tapia, Inmaculada Colmenar and José Albaladejo
Physical Chemistry Department, Faculty of Chemical Science and Technologies, University of Castilla La Mancha, Ciudad Real, Spain

Index